Let Nature Be Your Teacher

Let Nature Be Your Teacher

Integrating Nature-Based Learning in the Elementary Classroom

Louise Ammentorp & Helen M. Corveleyn

ROWMAN & LITTLEFIELD
Lanham • Boulder • New York • London

Published by Rowman & Littlefield
An imprint of The Rowman & Littlefield Publishing Group, Inc.
4501 Forbes Boulevard, Suite 200, Lanham, Maryland 20706
www.rowman.com

86-90 Paul Street, London EC2A 4NE

Copyright © 2025 by The Rowman & Littlefield Publishing Group, Inc.

All rights reserved. No part of this book may be reproduced in any form or by any electronic or mechanical means, including information storage and retrieval systems, without written permission from the publisher, except by a reviewer who may quote passages in a review.

British Library Cataloguing in Publication Information Available

Library of Congress Cataloging-in-Publication Data

Names: Ammentorp, Louise, 1972– author. | Corveleyn, Helen M., 1977– author.
Title: Let nature be your teacher : integrating nature-based learning in the elementary classroom / Louise Ammentorp & Helen M. Corveleyn.
Description: Lanham, Maryland : Rowman & Littlefield, [2024] | Includes bibliographical references and index.
Identifiers: LCCN 2024026913 (print) | LCCN 2024026914 (ebook) | ISBN 9781538161616 (cloth) | ISBN 9781538161623 (paperback) | ISBN 9781538161630 (epub)
Subjects: LCSH: Place-based education. | Nature—Study and teaching (Elementary) | Science—Correlation with content subjects. | Social learning. | Cognition in children.
Classification: LCC LC239 .A46 2024 (print) | LCC LC239 (ebook) | DDC 370.11/5—dc23/eng/20240628
LC record available at https://lccn.loc.gov/2024026913
LC ebook record available at https://lccn.loc.gov/2024026914

Contents

Preface	vii
Acknowledgments	ix
Introduction	1
1 Nature as a Healing Space	7
2 Nature and Language Arts: A Symbiotic Relationship	23
3 Schoolyard Dependables: Essentials for Outdoor Success	37
4 Agriculture, Gardens, and Native Plants—OH MY!	49
5 Hydroponics and Aquaponics: Clean, Green Growing Machines	69
6 A Healthy Harvest: Culinary Arts in the Classroom	87
7 Life Is Short, Play in the Dirt: Earth Science, Physical Science, Engineering, and Math	99
8 Rooted in Nature: Authentic Field Methods	115
9 Learning from the Past, Changing the Future: Social Studies, Advocacy, and Action	139
References	153
Index	157

Preface

Our Message: Let Yourself Bloom!

Our goal in writing this book is to provide educators with strategies and resources for bringing nature into the classroom and children outside. We are honored that you have chosen this book with the aim of bringing more nature-based learning to your students. Some of you may be doing a lot already and just want some new ideas; others of you may not have much experience bringing children outside for learning. During the Covid pandemic, in many districts there was a loosening of the rules about taking children outside for learning that have had long-lasting positive effects. However, there may still be resistance in some schools. Some people see obstacles in any new initiative, rather than potential. Some teachers and administrators (and students) may think of the outdoors as recess only. Some may say children can't focus outside, that there are too many distractions. Throughout our book we have included research on the benefits of outdoor learning as well as guidelines for successful learning. We encourage you to find other teachers in your school—grade teams, specials teachers, a PLC (Professional Learning Community)—who are willing to bring children outside for learning and bring nature into YOUR classroom.

We are delighted that you find value in looking for ways to incorporate nature into your teaching. Teachers play a vital role in nurturing young children's genuine connection with our planet and fostering their sense of environmental stewardship.

About the Authors

LOUISE AMMENTORP

Some people know they will be teachers from a young age. They play teacher at home and they dream of having their own classroom. This was not my path. I stumbled into my life as a teacher. I started out as a toddler and preschool teacher. I love the joy and wonder of this age group as they observe an insect busy at work, or as they jump into a puddle over and over again. Through my work with young children I held on to my sense of wonder and love of nature. I left the classroom and became a professor, primarily with preservice teachers in the elementary-school setting. After a few years, I became disheartened with the public education system: the worksheets, the pressure on teachers and students, the lack of opportunity for children to go outside and experience joy, which is so fundamental to the lives of children (and adults). I also saw the anxiety and stress experienced by preservice teachers who went through the school system. I questioned everything and considered leaving the world of education altogether.

Then I met Helen Corveleyn. I had two students placed in her classroom at Hopewell Elementary. I remember walking into her classroom the very first time. The room was bustling with energy and excitement. The fifth graders were creating models of machines that could clean the garbage patches in the ocean. I soon learned about all the initiatives Helen was leading at the school. She is a force of positive energy and great ideas. I felt hopeful and excited about teaching again and began to incorporate nature-based learning into my own life and work with preservice teachers. I began the journey of writing this book to share ideas and resources with other educators, with the hope that they, too, will feel inspired to bring the joy and wonder of nature into their own lives and the lives of the children they teach.

HELEN M. CORVELEYN

I began my teaching career as a seventh-grade life science teacher. My classroom was blessed with a door to an outdoor courtyard. I offered to all of my students the following opportunity: get to class early, and you can take a supervised run outside until I'm ready to start class. Word got out, and I had students that would tell me, "I've been watching the clock all morning, waiting until I could come and be outside for a little bit." On the days it rained they would literally show up with raincoats so they wouldn't miss the opportunity. It occurred to me that they were starved for outdoor access. I began integrating fieldwork into my lessons, even if it was just one aspect of a lesson that was taught outside. On these days, I rarely had any behavioral issues; students were completely engaged and learning was deep—not just surface-level vocabulary recall, but true *thinking*. I carried these experiences with me as I continued to teach through the seasons of my life, transitioning into a teacher-mom, and then returning to the classroom after raising my babies.

As I entered the realm of elementary-school teaching, expectations changed, behaviors changed, and attention spans definitely changed. I once again turned to outdoor learning as a way of connecting kids to nature and of connecting to the material they were learning inside the four walls of the classroom. When I met Louise, she was a constant source of compliments and positive energy. She elevated my voice and encouraged me to publish in the professional realm of teaching. Writing a book with her is one of the greatest privileges I've had in my career. I hope that this book will provide some ideas and some methods of delivering the beauty our Earth holds and the peace and exhilaration we can find in natural settings. Aside from the myriad of scientific evidence that exists about how important it is for us to be in nature, the simple truth is nature is brilliant. It has existed successfully since the beginning of time. It is our greatest and most wise teacher.

Acknowledgments

Thank you to everyone who contributed to the completion of this book. Your support and encouragement have been invaluable throughout this journey. In particular, we would like to thank the following teachers for their thoughtful contributions to this book. We are grateful for their inspired ideas!

Natasha Agrawal
Kate Atkins
Keeko de la Pena
Jeannine Lanphear
Naomi Montalvo
Ashley Thompson

FROM HELEN

This book has been a labor of love and passion that began when Louise walked through my classroom door and we immediately found a mutual respect and appreciation for one another's craft. I will always be grateful for our relationship and for her cheerleading skills, tropical vacation daydreaming, and endless GIF responses to my text messages.

To the students, families, and colleagues at HES who always welcomed me into their classrooms as "the Lettuce Lady," Mrs. STEM, and "Mother Nature," my memories of you are sweet, happy and lifelong. To Colby for our incredible work merging the beauty of poetry with the magic of nature. To Dr. Lauren, who inspires me as a scientist, mom, and friend. To Leah for her amazing artistic vision and collaboration that always turned out "just right." A special thank you to David for "always saying yes" to every idea and opportunity.

To my Project Dragonfly family at Miami University of Ohio, the EE friends I have made and learned from, and especially for the connection to the Vermilion Sea Institute in Baja, Mexico. To Meghann and Courtney for love, encouragement, and a window into nature unlike any other in my life. Parts of this book were written in Baja, while I was loved and lifted by the nature and people there.

To my three loves: Jonah for the smiles and warm hugs of love and support, Grace for fierce environmental companionship and planetary sisterhood, and Gabriel for all of the "you're doing a great job momma" encouragement.

To my family, especially my parents Nonny and Paul for developing my early appreciation for nature with my sisters Kate, Emily, and Maggie in national parks, scouting, and playing in the backyard. The core six were nature lovers from the "Clinkers" to Napa Valley!

To my husband Graig. My lifelong best friend, the one who believes in me before I believe in myself, and the love of my life. Thank you for nurturing this work right by my side.

I hope that you, reader, will find the joy, the exhilaration, and the humbling lessons that nature teaches us to guide your mindset as beautiful as I have.

FROM LOUISE

Thank you to my coauthor, Helen, for your inspiring ideas, your joyful spirit, and your friendship. You are truly a force of nature!

I am grateful to The College of New Jersey for granting my sabbatical leave, which enabled me to dedicate time and energy to the completion of this book. Thank you to Miriam, my sabbatical writing partner; your enthusiasm kept me motivated and inspired! I am also thankful to my amazing colleagues and students at TCNJ. Your support has meant so much to me.

Thank you to my family. My best memories as a child are playing outside with my sisters, Eva, Annette, and Tania. To Keith and Kaya, you are my everything. Thank you for letting me work when I have to and making me laugh when I need to!

Introduction

Come forth into the light of things,
Let Nature be your teacher.

She has a world of ready wealth,
Our minds and hearts to bless—
Spontaneous wisdom breathed by health,
Truth breathed by cheerfulness.

—Excerpt from "The Tables Turned" by William Wordsworth

The title of our book, *Let Nature Be Your Teacher*, draws inspiration from William Wordsworth's poem "The Tables Turned." Although written in 1798, Wordsworth's plea to learn from nature resonates in our modern context. Regrettably, children today spend significantly less time outdoors than any previous generation. Some blame technology for the lack of time children spend outside. Children 8–12 years old spend an average of four to six hours a day, and teens spend an average of nine hours, in front of electronic media (American Academy of Pediatrics Council on Communications and the Media, 2013; American Academy of Child and Adolescent Psychiatry, 2020; National Recreation and Park Association, 2017). The COVID-19 pandemic exacerbated this trend, with children as young as preschool spending up to seven to eight hours daily online for remote learning.

However, technology isn't the only reason children are spending less time outside. Other reasons include: less access to nature in urban areas; fear of letting children go outside unsupervised; the focus of academic performance in schools, leading to less time for play and recess (Charles & Louv, 2009; Louv, 2008). Education in the United States has increasingly become an indoor endeavor, perhaps reaching a peak in the past twenty years. In the early 2000s, No Child Left Behind ushered in an era of high-stakes testing, accountability, and a narrow view of academic achievement. Recess and outdoor time were cut, along with the arts, giving outcome-based language arts and mathematics instruction center stage. Although time for STEM (science, technology, engineering, and mathematics) instruction has increased in the past few years, it is not uncommon for young students to learn the parts of a plant through worksheets, photos, or videos, without venturing outside or even gazing out the window to observe plants growing in their natural environment. It is stunning that this is, unfortunately, the norm. Some reasons teachers give about why students are not taken outside for learning include: time and curriculum constraints; safety and class-management concerns; lack of administrative support; fluctuations in weather; and not knowing how to connect content with the natural environment (Dillon et al., 2016; Oberle et al., 2021). Nevertheless, there is a growing movement focused on bringing nature into the classroom and actively engaging children with the outdoors. This movement is gaining momentum as educators, parents, administrators, and researchers increasingly recognize the multitude of benefits associated with nature-based learning.

A Resurgence, Not a New Idea

Nature-based learning has deep historical roots that span across different cultures, religions, and educational philosophies. Indigenous cultures passed down knowledge of the natural environment through emphasizing the interconnectedness of all living things. These cultures recognize the invaluable lessons that the natural world offers, fostering a deep reverence for the delicate balance that sustains our planet. Within the Western intellectual tradition, early philosophers such as Aristotle emphasized the importance of observation and direct experience with nature as foundational to the learning process. In the eighteenth century, Jean-Jacques Rousseau's seminal work *Emile, or On Education* introduced the concept of "natural education," highlighting the importance of allowing children to learn through firsthand experiences in their natural environments. Rousseau believed that direct communion with nature cultivated curiosity, facilitated self-discovery, and nurtured moral virtues. Rousseau's ideas influenced later educational figures such as Jean Piaget and John Dewey.

In the nineteenth century, the "nature study" movement gained prominence, with educators like Louis Agassiz and Anna Comstock advocating for studying nature firsthand to foster curiosity and scientific observation skills in children. The Progressive Education movement of the early twentieth century advocated for an interdisciplinary curriculum grounded in student-driven inquiry and experiential learning. John Dewey, a leader of this movement, emphasized that education should possess real-life relevance and be firmly rooted in authentic experiences. He contended that students learned most effectively when actively engaged with their surroundings, encouraging inquiry-based learning wherein they explored and sought answers to their own questions. Dewey, alongside other progressive educators, criticized the rigidity of the indoor classroom and advocated for outdoor learning.

Despite the rich philosophical and programmatic tradition of nature-based educational approaches, mainstream public education has predominantly adopted an indoor-centric approach over the past century, reaching its height in the last two decades. The early 2000s ushered in an era marked by high-stakes testing, stringent accountability measures, and a narrow focus on academic achievement through initiatives like No Child Left Behind. Although initially on the fringes, the movement for nature-based learning has also grown. One of the major pioneers of this movement is Richard Louv. In his book *Last Child in the Woods: Saving Our Children from Nature-Deficit Disorder* (2005), Louv highlighted the benefits of nature for children's development and the negative consequences when we lack a connection to nature. The limitations and consequences of the "academic achievement" approach have become increasingly evident, particularly during the pandemic. Driven by concerns about disconnection from nature and a growing awareness of the educational and psychological benefits of the outdoors, immersive experiences such as outdoor and wilderness education, as well as forest schools and nature-based preschools, are increasingly popular. Organizations such as Children and Nature Network, the North American Association for Environmental Education (NAAEE), and its early-childhood counterpart, Natural Start Alliance, have brought nature-based learning from the fringes of education into the center.

Nature-Based Learning Organizations

- Children and Nature Network https://www.childrenandnature.org/
- Eco-Schools https://www.nwf.org/eco-schools-us
- Natural Start Alliance https://naturalstart.org/
- North American Association for Environmental Education (NAAEE) https://naaee.org/
- The Nature Conservancy www.natureworkseverywhere.org
- Project Learning Tree www.plt.org
- National Audubon Society's Audubon Adventures www.audubonadventures.org
- Outdoor Learning Store https://outdoorlearningstore.com/
- Center for Ecoliteracy www.ecoliteracy.org
- National Environmental Education Foundation https://www.neefusa.org/

What Is Nature-Based Learning?

Nature-based learning (NBL) includes both bringing nature into the classroom and learning in outdoor settings. As articulated by Cathy Jordan and Louise Chawla (2019):

> It encompasses the acquisition of knowledge, skills, values, attitudes, and behaviors in realms including, but not limited to, academic achievement, personal development, and environmental stewardship. It includes learning about the natural world, but extends to engagement in any subject, skill, or interest while in natural surroundings. (p. 2)

Nature-based learning isn't confined to science; it surpasses disciplinary boundaries, creating an environment conducive to more authentic and holistic learning experiences. A mounting body of research underscores the advantages of this approach. In the article titled "Do Experiences with Nature Promote Learning? Converging Evidence of a Cause-and-Effect Relationship" (Kuo, Barnes, & Jordan, 2019), the authors conduct a comprehensive review of research findings, providing compelling evidence indicating that nature-based experiences significantly enhance academic learning, personal development, and nurture a sense of environmental stewardship.

4 INTRODUCTION

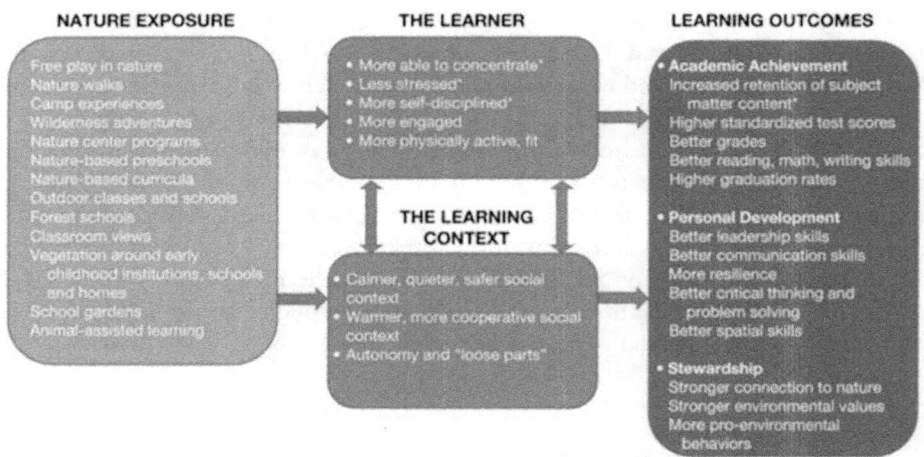

Figure I.1. The Benefits of Nature-Based Learning (Copyright © 2019 Kuo, Barnes, and Jordan)

Figure I.1, reproduced from the article, provides a helpful overview of the varied ways children can be exposed to nature, the impact of nature on the learning context, and the benefits to the learner, socially, emotionally, academically, and physically. It also highlights the pro-environmental behaviors and values that students demonstrate when they develop a deeper connection to nature. However, there is much work to be done to bring nature-based learning to all students, particularly in the public school context.

The Structure of the Book

Our book highlights the importance of, and strategies for, integrating nature-based learning into the elementary-school curriculum. This includes bringing children outside for learning and bringing nature into the classroom. Our goal is to offer ideas that provide benefits to student academic learning and positive mental health. Each of our chapters offers ideas that educators can immediately apply, no matter the budget or location. We provide a selection of lessons and activities that can be used by both new and experienced educators.

Overview of Chapters and Descriptions

CHAPTER 1: NATURE AS A HEALING SPACE

Often people think of the outdoors as a place for recess, where children can be loud and physically active, as a "break" where they get their energy out. This chapter highlights the ways nature can create a calming space for students to reset, refocus, and relax. We discuss how nature-based learning supports students' development of Social-

Emotional Learning (SEL) competencies. We provide ideas for several short, accessible SEL activities such as yoga stretches and mindfulness exercises.

CHAPTER 2: NATURE AND LANGUAGE ARTS: A SYMBIOTIC RELATIONSHIP

This chapter focuses on the many ways to bring nature into the classroom through language arts. We provide ideas for supporting writing instruction through nature-based materials and going outside. We highlight a unit created in partnership with a visiting poet that includes outdoor nature-based poetry lessons that can be done as part of a full unit or individual lessons.

CHAPTER 3: SCHOOLYARD DEPENDABLES: ESSENTIALS FOR OUTDOOR SUCCESS

Nature can be found in all settings—whether a school has a blacktop playground, fields of green, snow, or sandy soil. We include lesson and activity ideas using nature found in the local environment, including birds, nests, and insects. We also include methods for students to actively observe and study nature ecosystems.

CHAPTER 4: AGRICULTURE, GARDENS, AND NATIVE PLANTS—OH MY!

Chapter 4 explores the many ways teachers can incorporate agriculture and gardening into their curriculum. Drawing on the National Center for Agricultural Literacy (2013) learning outcomes for agricultural literacy, we focus on the themes of *Plants and Animals for Food, Fiber, and Energy; and Food, Health, and Lifestyle*, and, in particular, fiber and gardens. From vegetable gardens, containers, and meadows to native plants, pollinators, and themed gardens, there are so many possibilities for teaching with children.

CHAPTER 5: HYDROPONICS AND AQUAPONICS: CLEAN, GREEN GROWING MACHINES

Hydroponics are a wonderful, hands-on method of looking at plant growth at every level of elementary education. Teachers can use hydroponic and aquaponic systems year-round and indoors, so they're an accessible way for educators to teach about plants and nutrition. This chapter will offer suggestions about how to get started and how to expand a garden, from tabletop to indoor farm. The chapter will also include Next Generation Science Standards (NGSS) aligned lessons and ideas for schoolwide programs.

CHAPTER 6: A HEALTHY HARVEST: CULINARY ARTS IN THE CLASSROOM

Using culinary arts and sciences in the classroom provides many different learning opportunities for students and teachers alike. Students love to eat what they cook. This chapter will provide lesson and activity ideas that integrate math, science, literacy, social studies, and SEL.

CHAPTER 7: LIFE IS SHORT, PLAY IN THE DIRT: EARTH SCIENCE, PHYSICAL SCIENCE, ENGINEERING, AND MATH

Teaching STEM in and with nature engages students in hands-on, experiential learning while fostering a deeper appreciation for nature. In this chapter, we look at ways to integrate nature into STEM teaching, with a focus on engineering, physical science, earth science, and math. A variety of lesson ideas for using natural materials for hands-on projects and activities, as well as ideas for bringing teaching outside, are included.

CHAPTER 8: ROOTED IN NATURE: AUTHENTIC FIELD METHODS

Providing tools that students can use to measure and investigate ecological relationships ignites the desire to gather evidence from the field. This chapter will give ideas and an in-depth example of a unit that creates authentic field experiences with students using tools and methods from the fields of conservation and biological and ecological sciences.

CHAPTER 9: LEARNING FROM THE PAST, CHANGING THE FUTURE: SOCIAL STUDIES, ADVOCACY, AND ACTION

This chapter offers nature-based meaningful lessons and activities for teaching history, geography, and mapping. We describe ways to encourage students to think about, and get involved in, important local and global issues such as climate change and environmental sustainability.

CHAPTER 1

Nature as a Healing Space

In every walk with nature one receives far more than he seeks.
—John Muir

Often when people think about children in the outdoors, loud and physically active experiences come to mind. In schools, the outdoors is often viewed as a respite from learning, an opportunity to get energy out through running around. While these are important experiences, they are not the only way children, and humans in general, experience the natural world. For both adults and children, the outdoors can also be a contemplative, quiet place—a place to observe, listen, and become centered. *The outdoors is both a place to get energy out as well as a place to get energy in.* Opportunities to experience nature in this way should be given to children during school hours.

This chapter explores the ways that teachers can support Social-Emotional Learning (SEL) through nature-based activities such as mindfulness and yoga. Simple techniques incorporated into the day provide students something to look forward to and connect students to nature. Positive experiences in nature as a young child cultivate empathy and a sense of environmental stewardship that will continue into adulthood.

Social-Emotional Learning

Over the past few years there has been a growing recognition of the need for SEL in schools and classrooms. Many states, such as New Jersey, have adopted SEL Competencies. The Collaborative for Academic, Social, and Emotional Learning (CASEL) Framework is one of the most comprehensive and utilized approaches to SEL learning. The CASEL Framework outlines the five core competence areas of SEL: self-awareness, self-management, social awareness, relationship skills, and responsible decision-making. These competencies underlie academic and professional success as well as health and well-being. Relationships are central to the development of CASEL 5. "SEL instruction is carried out most effectively in nurturing, safe environments characterized by positive, caring relationships among students and teachers," according to the CASEL website.

> **Teacher Resources**
>
> **Books**
>
> - *Mindfulness & Nature-Based Therapeutic Techniques for Children: Creative Activities for Emotion Regulation, Resilience and Connectedness* by Cheryl Fisher
> - *Naturally Inclusive: Engaging Children of All Abilities Outdoors* by Ruth Wilson
> - *The Curious Nature Guide: Explore the Natural Wonders All Around You* by Clare Walker Leslie
> - *The Nature Connection: An Outdoor Workbook for Kids, Families, and Classrooms* by Clare Walker Leslie
> - *Growing Your Own Tea Garden* by Jodi Helmer
> - *Herbal Tea Gardens* by Marietta Marshall Marcin
> - *Yoga for Children: 200+ Yoga Poses, Breathing Exercises, and Meditations for Healthier, Happier, More Resilient Children* by Lisa Flynn
> - *Yoga Pretzels: 50 Fun Yoga Activities for Kids & Grownups Cards* by Tara Guber, Leah Kalish, and Sophie Fatus
>
> **Websites**
>
> - Collaborative for Academic, Social, and Emotional Learning (CASEL) https://casel.org/fundamentals-of-sel/
> - SEL4US https://sel4us.org/
> - Choose Love Movement https://legacy.chooselovemovement.org/
> - Three ideas for taking SEL outdoors https://medium.com/edumotion-sel/3-ideas-for-taking-sel-outdoors-962cfa126bf2
> - Mindful Schools https://www.mindfulschools.org
> - Yoga 4 Classrooms https://yoga4classrooms.com/

CASEL emphasizes that a student's sense of belonging is essential to student engagement and a strong school culture.

There is an ample amount of research on the long-lasting academic and behavioral benefits of SEL instruction (Durlak et al., 2011; Taylor et al., 2017). Further, research shows that the most effective SEL programs are those run by teachers (Durlak et al., 2011). There is also considerable research on the physical and mental health benefits of the outdoors for children and adults. For example, Tillmann et al. (2018) conducted research that explored the impact of natural spaces on individuals' emotions, specifically focusing on the experience of awe. Awe is an intense emotional response characterized by a sense of wonder, astonishment, and reverence often triggered by encountering something vast, transcendent, or awe-inspiring. Tillmann's findings support the notion that natural spaces have the ability to evoke a sense of awe and positive emotion. The experience of awe in natural settings was associated with positive emotions such as joy, contentment, and a sense of connection to something greater than oneself. Research conducted by Ballew and Omoto (2018) also explored the relationship between natural spaces, awe, and positive emotion. Their study examined the impact of engaging with nature on individuals' well-being and happiness. They found that spending time in natural environments was associated with increased positive emotions, including awe, gratitude, and a sense of beauty. These findings underscore the importance of incorporating nature experiences into children's and adults' lives and recognizing the positive impact they can have on emotional well-being.

Although the CASEL Framework does not mention the relationship to the natural world, nature functions as a fertile ground for children to grow in each SEL competency area. The natural environment is a powerful and important context for developing a sense of belonging and connectedness to the world. Nature is not just the backdrop or setting for social relationships; instead, relationship with nature supports the development of SEL competencies.

One of the true benefits of practicing SEL in nature is that children naturally "soften" when they go outside. The rigidity of the classroom space is dissolved and kids are instantly more open to the message of mindfulness. It's like thinking of a school of fish in an aquarium. The fish in the tank can be happy swimming around together in the aquarium for a long time, but if you let them out into the ocean they are no longer confined by the glass of the aquarium; they are in water that is no longer artificially controlled and natural sunlight allows them to see the ocean in a whole new way.

Photo 1.1. Hopewell Elementary School (HES) Infinity Garden (Helen Corveleyn)

Stress, Mindfulness, and Nature

Unfortunately, stress and anxiety are becoming increasingly common among children and adolescents. Academic-related stress is a significant contributing factor to anxiety in children. According to the National Institute of Mental Health (NIMH), approximately 31.9% of adolescents aged 13–18 in the United States experience an anxiety disorder at some point in their lifetime. These statistics were recorded before the COVID-19 pandemic, which has had a profound impact on children's mental health, leading to increased levels of stress and anxiety. One study revealed that 71% of parents reported that the pandemic had negatively affected their child's mental health (Abramson, 2022). Furthermore, the Centers for Disease Control and Prevention (CDC) documented a notable rise in mental health–related emergency department visits among children and adolescents during the pandemic. Since 2019 mental health–related emergency room visits for children aged 5–11 have increased by approximately 24% (Leeb et al., 2020) This alarming trend underscores the urgency of addressing the mental health challenges facing young people.

In recognition of the stress and anxiety faced by students, it has become increasingly common for schools to integrate mindfulness activities into the classroom. Often this includes a short video in between lessons or sitting quietly. However, mindfulness is more than an activity; it embodies a mindset and classroom culture that needs to be practiced and revisited. Mindfulness is about intentionally bringing one's attention to the present moment with an attitude of curiosity, acceptance, and nonjudgment. It involves being fully aware of thoughts, feelings, bodily sensations, and the surrounding environment. It encourages a deeper connection with oneself, others, and the world, fostering a greater sense of clarity, compassion, and overall mental well-being. It includes practices for settling, centering, or increasing focus.

Looking at Life through Younger Eyes: Mindfulness Teaching Refreshes the Teacher's Soul

(Helen)

Sitting under the canopy of branches, children close their eyes and dig their heels into the blanket of pine needles. "Stomp your heels, and take a deep breath in. What do you smell?" "Christmas," they answer.

Outside with kindergartners. The rain is slowly falling through the boughs of the trees overhead, our eyes are closed. As the rain falls, the kids say, "Let's listen to what the rain is saying to us." Encouraged and surprised at the depth of their intensity of the meditation, we sit for ninety seconds (a long time for kindergarten) and their bodies are totally still. When we open our eyes, I ask them what the rain said to them. One immediately raised his hand and looked deeply into my eyes and said, "That rain said it just wants to get into my SHIRT!" Giggling, I called it a "win."

When you teach mindfulness to our youngest learners, you never know what type of "unfiltered" viewpoints they will offer you, which makes it amazing to teach. When my preschoolers told me that the air smelled like Christmas after stomping on pine needles—in the middle of the heat of New Jersey late-spring temperatures—it was a reminder of the joys of being young. Often I walk away feeling mentally lighter because teaching young children reminds me of the beauty of looking at life simply.

Nature-based mindfulness deepens our appreciation of nature, fosters a sense of interconnectedness, and provides a space for personal growth and well-being. It allows us to slow down, engage our senses, and develop a profound relationship with the natural world around us. Students will feel a deeper impact in participating in outdoor mindfulness activities due to the combination of fresh air, natural sunlight, and a calming effect of green spaces.

Incorporating Nature-Based Mindfulness in the Classroom

Mindfulness can be practiced in silence or with quiet words, individually or in groups. A key aspect of practicing mindfulness is that the teacher must be engaged. If the teacher is asking students to close their eyes or lower their eyes to the ground, they should model it. The more mindfulness is integrated into teaching, the more the children look forward to it. This can then translate into better classroom management as students practice SEL competencies such as self-awareness and self-regulation. It also translates into a more relaxing and focused classroom community. Mindfulness works best when embedded in the routine throughout the day. It includes noticing and discussing sensory experiences, as well as articulating a self-awareness of sensory and emotional experiences.

CALM BODY, CALM MIND

Silent mindfulness can be practiced as an individual relaxation technique paired with breathing and anxiety reduction. Silence in a natural setting can have therapeutic benefits (Pfeifer, Fiedler, & Wittmann, 2020). When first incorporating mindfulness in the classroom, sometimes educators will say, "This is a great idea but this group cannot handle that." Or, "I can appreciate the importance of this skill but they just can't sit still." It is important to note that even a minute of mindfulness does not need to happen right away. Just like exercise, mindfulness is a skill that needs to be practiced and stamina needs to be built. For example, preschoolers may only be able to sit still for thirty seconds. The next time they try being mindful, they can work up to a minute, three minutes, and beyond.

PRE-TEST BOOST

Oxygen and sunlight outside release serotonin, which regulates mood and helps the brain to focus. For a pre-test boost, take kids outside. Have them stand in a circle and offer a compliment to each classmate. A few deep-breathing techniques and some stretching, and then re-entry into the classroom for a focused activity like a discussion or assessment: the teacher will see a significant difference in performance with as little as a ten-minute investment.

A MINDFUL MINUTE

One minute of stillness can precede a mini-lesson, an assessment, or a writing assignment that needs quiet, focused attention.

- **Set a timer** (this is important for kids to see so they know you are conscious of the time).
- **Set your outdoor focus.** Ideas could include: the warmth of the sun, the feeling of grass when you run your fingers over it, the smell of rain coming, feeling a breeze on your skin, the sound of birdsong.
- **Engage one of your senses other than eyesight.** Encourage kids to engage one sense other than eyesight to become aware of the world around them. Because people often feel uncomfortable closing their eyes in a group, offer something to blur their focus other than just closing their eyes. Students might feel comfortable picking one spot on the ground to focus on and relaxing their eyes on that fixed point while letting their surroundings fade around them to allow the chosen sense to be dominant.
- **Get set.** Always begin by asking students to be in a comfortable position, relax their body, and let distractions go. Take a few deep breaths (the more you practice mindfulness, the more you can incorporate breathing).
- **Be mindful!** Students will often ask for more after sixty seconds. You know your classroom best, so you can determine whether you want to leave them wanting more, or attempt another experience after the first minute. The more frequently you practice this, the more the children will feel comfortable and seek out this practice.

GUIDED VISUALIZATION

If you have a population that looks forward to mindfulness time, you can build their endurance and take longer periods to be mindful. Begin to stretch student thinking beyond a one-minute limit. Following your recess time, make mindfulness a priority and link the practice to a time that you are already outside. The kids are dressed in their outdoor clothing and have just exercised. Before returning inside for other periods of learning, teach them to focus and reset to prepare their brains for learning. If students are practiced in five- to ten-minute periods of mindfulness, when they have endurance to be still, introduce methods of reducing anxiety.

> **Peaceful Prompts**
>
> When you deliver a mindfulness prompt, keep your voice calm and steady. Your voice should sound different from your "teacher" voice. You can even say, "When you hear my voice change from my teacher voice to my peaceful voice . . ." Make sure to pause between each statement so students can feel the change from an instructive exercise to a relaxing exercise. Here is an example:

> *I know we are all preparing to take our assessment in math next period. A great way for us to prepare for our assessment is to let our brain open and clear away anything that might prevent us from remembering our math. Sit with your legs and hands in a relaxed position. Take five rainbow breaths and feel the cool air that surrounds you. Picture all of the peacefulness of nature. Now choose something in nature that you love. It could be an animal, the ocean, a walk in the forest . . . picture that part of nature cheering you on and encouraging you because it knows you are ready to take your math test. Now picture yourself being confident, focused, and knowledgeable. Feel the solid ground where you sit, and gather strength from the Earth. Take that strong, stable power and bring it in with you to your math test. You feel energized and focused.*

A guided meditation may sound like this:

> *Imagine being in the midst of the vast ocean and hearing the distinctive sound of a whale breath. As your boat slows, the occupants begin to pinpoint the area where the breath came from. You get a glimpse of the dark, shining back of a whale. Minutes later a few more glimpses. Your heart begins to beat a bit faster, and you have an overwhelming sense of excitement and hope that a bit more of the whale will reveal itself: you are rewarded with the full view of its tail, or perhaps if you are truly lucky, for the whale to breach in front of you. The feeling following this encounter is unmatched by anything else. Because it is fast, unexpected, and fleeting, your brain takes a little while to connect with your emotions.*

PHOTOGRAPHY

A recent study (Jo, Song, & Miyazaki, 2019) from the National Institutes of Health found that even looking at photographs of natural areas significantly reduces psychological stress. Further, looking at green foliage, flowers, and wooden materials contributes to positive effects on the prefrontal cortex. There are wonderful photography books to have in the classroom library, such as those by Andy Goldsworthy. Having students take pictures of nature and making a class book and/or posting photos around the classroom are great ways to integrate photography.

FOUND OBJECTS AND NATURAL MATERIALS

Mindful experiences need not be individual sedentary activities. Quiet group mindfulness is when students can be engaged in a mindful activity that is repetitive and calming: this could include arranging found objects in nature in a shape or design, arranging rocks in patterns to create a path, or engaging in yoga poses as a class outside. These physical activities are useful particularly for young children who may need concrete objects to manipulate for focusing their attention.

LEAF BOUQUETS

Bring students to an area where leaves have fallen from trees in autumn. Begin with a few big breaths and a "gratitude thought" about change. Then explain how the leaves change in the fall as the trees sense winter coming. Tell the students to look and identify leaves that show beauty. If you are in an area with a lot of leaves, the tendency of this age group is to gather up large armloads of leaves to get the "most." The mindful activity is to pick them up one at a time and create a bouquet in their hand. The bouquet requires some hand strength and coordination, particularly for young students.

LOOKING FOR *NATURE'S SURPRISES*

In her books, Clare Walker Leslie discusses the simple act of having students go for a walk and find surprises in nature. This is a wonderful activity for building observation skills and encouraging a mindset of gratitude and appreciation for the beauty of the natural world, no matter the context. Look up at the sky! The shape of a cloud could become a wonderful surprise that evokes creativity, laughter, and class camaraderie. If necessary, walk a few steps and model finding a surprise for your students. It can be a bird's nest hidden in a bush, the first sign of flowers in springtime, or even something silly (a pile of deer droppings in the middle of a path—poop always gets a good giggle!).

Photo 1.2. Nature Mandala (Helen Corveleyn)

COLLAGES, MANDALAS, AND SPIRALS

Collages and mandalas can be exciting for every age. The idea of hunting or collecting is a thrill for all students because who doesn't like to gather and then come back to an area to build? This is a wonderful way of connecting beauty to creation. Mandala means "circle" in Sanskrit. Mandalas are used for intentional mindfulness and to bring healing to the soul. You may feel comfortable sharing this with students or you may say that mindful actions cause us to slow down, recognize the beauty in our surroundings and in ourselves, and what is left behind is a gift for anyone walking by to enjoy. Students may need to outline an area first to get started, as not everyone feels comfortable with a blank canvas to begin. We frequently find two branches that fit together in a unique shape to begin to build our mandala. One or two students stay to arrange the objects, and the rest in the group gather flowers, seeds, leaves, sticks, woodchips, or acorns to create the inside concentric rings.

Mindful Hard Work

The feeling of accomplishment from doing hard work and helping others cannot be underestimated. The labor of gardening can be physically demanding and repetitive, which is often what students need. Physical activity is good for children: it releases endorphins and decreases depression. Helen shares her experience working with the occupational therapist at her school to support a student:

> I had a student who hated going to occupational therapy. The child was happy, bright, and energetic, but when it came time for OT, they groaned and complained and turned into a completely different student. I consulted with the OT and asked why this was happening. They said the student has a lot of weakness in their forearms, wrists, and hands that contributed to their dysgraphia. The child, however, loved being in the garden. We set to work on a hybrid schedule where the student had extra time in the garden, pulling weeds and using the trowel and hand cultivator to turn soil to provide an incentive for the indoor OT that needed to happen, and the gardening helped strengthen the muscles needed in their arms at the same time. It was a win-win for both the garden and the student!

There are many ways that nature-based activities are beneficial for children with special needs. *Naturally Inclusive: Engaging Children of All Abilities Outdoors* (2023) by Ruth Wilson is a wonderful resource for educators.

The Power of Service and Reflection

As part of the work of the Green Team at Hopewell Elementary School (HES), students conduct a trash collection around our school grounds. With the addition of outdoor pandemic-style eating (with all snacking done outside), the increase of litter was remarkable. As we walked the grounds picking up litter, the students commented that the searching for and picking up trash felt mindful. After the walk they reflected on how, even though they were doing something "dirty," they were thinking about the care of the Earth and noticing the improvement of the world around them. Their comments illustrate the power and impact of creating a culture of mindfulness through nature-based service opportunities. It also points to the importance of creating the time and space for reflection. Reflection on their actions supports self-awareness, social awareness, and nature awareness.

Often as teachers we notice these benefits and can talk about it with colleagues or through lesson-plan reflections, however, it is important to have these discussions with students. Prompts can include, "How did you feel mindful when we picked up trash?" or "What feelings did cleaning-up elicit from the start compared to how you felt when we finished?" Drawing their attention to being mindful also helps them to self-identify mindful activities in the future and supports pro-environmental behaviors.

Creating Community and Calm through Teatime

There are few things better than warm tea as a way to calm down and also to create a special time. However, teatime is rarely seen in the classroom. One obvious obstacle that people may initially think of is that hot water and children don't mix well. Clearly, the teacher should be careful when handling hot water. However, that doesn't have to be a deterrent. The water can be heated out of the reach of young children. The water can be boiled and the tea can be placed in a thermos until ready for consumption. If you have a particularly challenging class, go for tea—but make it iced! It is best if children can assist with dried herbs and observe the steeping process, if possible. Tea sets, mugs, and trays can be donated from the community or purchased at a secondhand store or garage sale. When possible, use herbs from the outdoor garden or hydroponic garden or even foraged.

There are many ways to integrate teatime into the classroom. It can be part of a daily routine in the class, a special occasion, or as part of a study.

- During snack time
- On cold days after coming in from outside
- While outside
- During meeting time
- As part of a study of different cultures, tea ceremonies, agriculture, and history of, for example, China, Japan, Morocco, India, or England.
- Exploring senses—it's a great way to describe observations of all five senses
- Connecting to literature/children's books

> ### Supporting Risk-Taking and an Explorer Mindset
>
> After working with National Geographic and Lindblad Expeditions during a Grosvenor Teacher Fellowship, Helen became acquainted with the "explorer mindset" as defined by the National Geographic Society. They have defined attributes in the categories of attitude, skills, and knowledge to create more empathetic, curious, and empowered learners as travelers of our world. The beauty of this model is that this mindset does not have to happen after you take your class on a long plane ride to an exotic location. It can happen in a contained space in a park near your school, in the woods behind the school, or even in a vacant lot near your location. As long as you put safety measures in place first, kids can explore in their own way to satisfy their curiosity at their own level. They may take risks and this is excellent as we want them to be risk takers and stretch themselves! As a teacher, this requires a "why not" attitude and deciding when risk is possible. For example, if a timid student is taking a risk by crossing a creek on a log that is sturdy (does not pose a safety risk) but they might step in a big puddle of mud at the end, it may be worth it to let them try. Rinsing off the shoe might not be such a big deal compared to the risk they took to step out of their comfort zone. The reward of the student feeling empowered to make their own decision to jump over the creek may far outweigh the cleanup of the mud! Of course, buy-in by the school and families is important for this kind of risk to take place.
>
> In collaboration with the school guidance counselor, Helen's students explored the wild space behind the school for fifteen minutes. They had no other limiters other than the fact that they were allowed to make up their mind about what they wanted to do. Between the grassy area of the school's backyard and the wild space, there is a small creek that students had to get over in order to get to a grassy meadow area. Students explored the area by themselves. At all times teachers could hear the students. This particular set of students had an understanding of limits and norms within the classroom. Here are just a few student quotes heard during the experience:
>
> - "We have to go on a log; man, this is like an adventure!"
> - "Be careful of thorns!"
> - "Will there be poison ivy? Tell me what poison ivy looks like so I know."
> - "I really want to get to the other side but I'm nervous about crossing the log."
> - "There are thorns in my boots and it hurts! But I'm cool with it."
> - "I found green scat!"
>
> The benefits of this type of interaction are many. Students are more self-aware, agentive, confident, and connected to nature. Whereas not every school has access to such a space, trips to local and state parks, environmental centers, and other outdoor areas can provide such opportunities to students.

Yoga in the Classroom

Kids love yoga! It's an organized activity that is physical, but also relaxing. Students know they can expect the teacher to also be relaxed, calm, and peaceful. It is amazing how much students respond to an activity that they know will be "Zen." There are many resources out there for kids' yoga, and the most important thing for beginner yoga leaders is to start small and easy. Yoga card decks are fun ways to explore if the teacher does not have yoga experience. Sometimes the beauty and camaraderie come from trying to figure out the poses *together*—it's a shared experience that is physical and can be easily modified for any learner.

Especially for younger students, the teacher may have to narrate the expected behaviors at first. The teacher may start by saying:

> *In order to have a relaxing experience today we are going to practice several things with calm bodies: Each person is going to stay in their own space [towel, mat, carpet square]. Everyone's talkers are on a break for a little while so you can hear my voice and the voice of nature. Our bodies need to act calmly, no wild arms or legs, just peaceful, slow movements. [This could also be practiced a few times.]*

Older students respond well with a brief centering breathing exercise and providing choice at the beginning, for example saying, "Would you all prefer to be in a circle, or should we create three lines?" Even small areas of control help kids feel more at ease if it is new to them. Throughout the exercise, always give kids an opportunity to stretch or position themselves at their own comfort level. Saying things such as, "If you'd like to try this pose and push through being uncomfortable, bravo for you! If it feels like too much for you, here's a modified version." There are many poses that connect to nature and animals, such as: Mountain Pose, Tree Pose, Eagle, Fish, Cobra, Cat. If it seems to be going well, and the kids are into it, a fun extension for kids is to pair up and create their own yoga poses based on nature.

Teacher in the Limelight: Schoolwide Therapy-Dog Program

In this section, teacher Ashley Thompson describes the program Tail Waggin' Tutors, which she created at her school.

For the last fourteen years, I have had the joy of teaching kindergarten at West Belmar School, located in Wall Township, New Jersey. West Belmar is a quaint K–5 school nestled in Wall Township, with a close-knit community of approximately 150 students. I am deeply passionate about guiding young learners on their journey to reading, equipping them with the foundational skills necessary for reading success. Fueled by extensive research and my own profound interest, I launched a unique animal-assisted reading initiative called Tail Waggin' Tutors at West Belmar School nearly thirteen years ago. This innovative program has been a resounding success ever since its adoption and continues to thrive to this day.

Since I became a teacher, I have always harbored a keen interest in introducing a dog-assisted literacy program in my elementary school. Over the years at my school, I observed challenges with student behavior and challenges with reading—students struggled to have an interest in reading, experienced difficulty with reading-fluency rates, and lacked confidence in their reading abilities. After thorough research and the successful composition of a school district grant, my proposal to introduce a dog-assisted literacy program gained approval. The program has been in effect for the

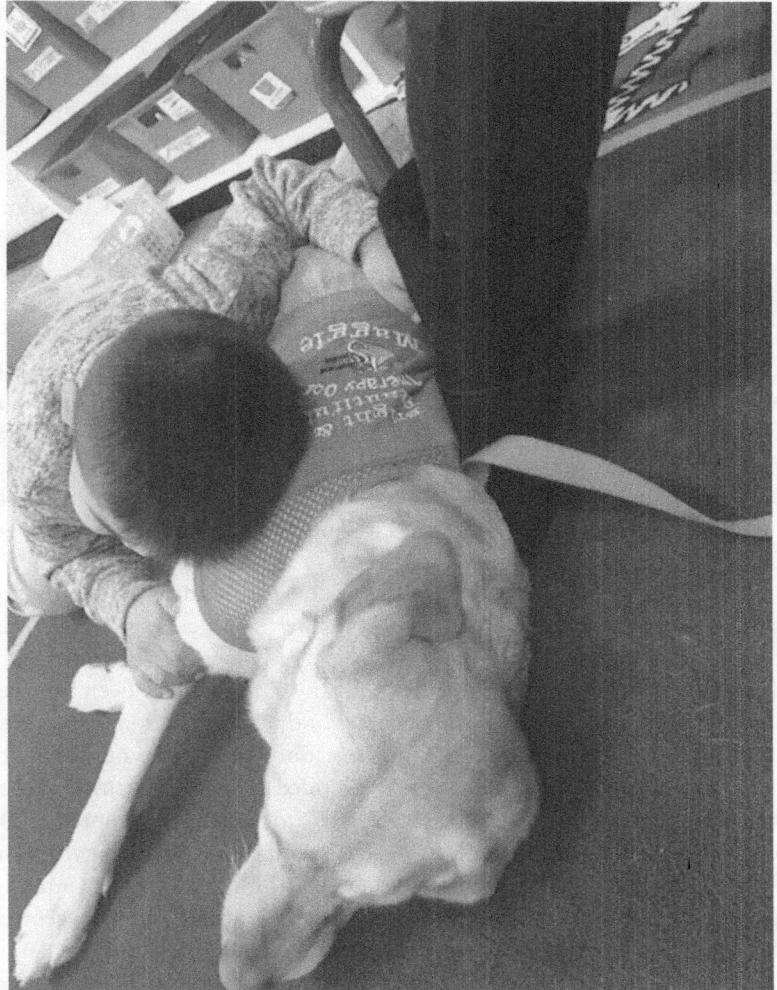

Photo 1.3. Canine Companion (Ashley Thompson)

past thirteen years. The grant outlined the objective of creating a relaxed and "dog-friendly" environment, fostering opportunities for students to hone their reading skills, bolster confidence, take risks, enhance fluency, and promote socialization—all in the comforting presence of a nonjudgmental canine companion. The initiative was structured to involve a licensed therapy dog visiting each classroom once a week during the students' designated reading block. The therapy dog and its owner functioned as a literacy center, where students engaged in reading books they were currently reading in class.

To enlist volunteers, I reached out to two local organizations specializing in certifying therapy dogs, namely, Therapy Dog International and Bright & Beautiful Dogs. These organizations, whose availability may vary by region (refer to https://www.akc.org/sports/title-recognition-program/therapy-dog-program/therapy-dog-organizations/ and http://landofpuregold.com/rxb.htm for a list of similar programs), were instrumental in spreading the word about this new opportunity. Certification requirements mandated that both the dog and handler pass tests assessing sociability, obedience, and adaptability to varied environments. Additionally, volunteers were required to maintain liability coverage and ensure the ongoing health of their animals through yearly renewals, safeguarding the well-being of both dogs and children.

Dogs and their handlers were invited to interview sessions with the school administration. During these sessions, owners shared their experiences with children and therapy-dog programs, provided references, communicated availability and grade-level preferences, presented certifications and insurances, and articulated their motivations for joining our program. The interviews concluded with select students interacting with the dogs and their owners, offering valuable insights into how these pairs related to children. Once all interviews were completed, we strategically matched each dog and handler with a specific classroom. The dogs varied from terriers to golden retrievers to Great Danes. Since the program is now in its thirteenth year, several dogs from the original group have passed away, and only one of the initial handlers remains an active volunteer, underscoring the ongoing difficulty of maintaining a comprehensive school program. This year we recruited a few more dogs and have our entire building covered, totaling ten therapy dog and handler teams.

In order to keep the program in place each year, students are provided with permission slips that necessitate signatures from their guardians, granting consent to participate in the program. The Wall Township Public Schools Board of Education granted approval for the program initially, and this approval has been renewed each year. Handlers diligently provide updated copies of their certifications each year. Upon the program's inception, I conducted a presentation to all staff, outlining the program's expectations, goals, and objectives. I provided numerous examples illustrating the diverse ways therapy dogs could be integrated into classrooms, tailored to different grade-level needs. I extended invitations to staff members interested in witnessing the program in action, encouraging them to observe how I incorporate dogs into my classroom. Most staff members have implemented the program as a literacy center where students read their weekly stories and texts to the therapy dogs and their handlers.

Before introducing our four-legged friends into the classroom, teachers engaged in conversations with their students about expectations and appropriate behaviors during the dogs' visits. Typically, teachers explain what a therapy dog is and its role in society and in the classroom, emphasizing its positive impact on SEL, boosting confidence in struggling readers, and enhancing overall reading skills through interactions with a nonjudgmental furry companion. Teachers ensure that students understand the importance of asking permission before petting the therapy dog and the need to refrain from getting too close to its face. Students are reminded to be gentle and to use indoor voices, recognizing that some dogs may be sensitive to noise or movement. During the initial visit, teachers organize a "meet and greet" session with the furry friend and its handler. This provides an opportunity for students to learn about their classroom therapy dog and its handler. Students are actively encouraged to ask questions, fostering familiarity with these regular visitors who become integral parts of their classroom environment.

During weekly visits, dog handlers ask students about their preferences—whether they would like to pet their dog or read to their dog. While the majority of students are enthusiastic about engaging, there are occasional instances where some may not be as willing. Factors such as a previous negative experience with a dog outside of school, lack of exposure to dogs, or simply having a challenging day might contribute to their hesitancy. Typically students who are hesitant in the beginning are usually excited to see the therapy dog after a visit or two.

Over the past thirteen years, I and all program participants have witnessed the role therapy dogs have played in managing students' emotions within the school building. Students encountering difficulties in their academic tasks have been motivated by the prospect of reading with a therapy dog or spending a break with the dog. We have used this program with our emotionally regulated students as well. The dogs have been used as a break for the students and this has served as a rewarding experience, allowing them to sit and share quiet moments with the dog before returning to their academic responsibilities. Enthusiasm fills the air as reading-assistance dogs make their way to West Belmar Elementary School, capturing the excitement of dogs, handlers, teachers, students, and staff alike. What sets the dog literacy program at this school apart is its distinctiveness—human-dog handler teams visit every class. Unlike typical programs that concentrate on aiding emerging and struggling readers, this inclusive approach spans the entire school.

Our schoolwide therapy-dog program has not only impacted student reading outcomes but it has also significantly contributed to Social-Emotional Learning (SEL) initiatives in several significant ways. Interacting weekly with therapy dogs and their handlers has helped students develop empathy. Students deeply care for and understand the needs of dogs, which we have seen translate into improved empathy and compassion toward their peers. Students who experience levels of stress and anxiety are much more relaxed in the presence of a therapy dog. For students who are experiencing stress and anxiety, the program makes it easier for them to engage in SEL opportunities as well as reading. Students who participate in the program are extremely excited about their weekly dog visits and you can see students' moods change when they know the dog is visiting. I have seen firsthand the simple act of petting a dog boosting a child's mood and providing comfort, which in turn enhances a student's readiness to learn and interact socially.

Over the years, the students have built relationships with the dog and their handlers. When therapy dogs visit, students must be cooperative and understanding. Learning to interpret and understand a dog's body language and respond appropriately can also enhance students' ability to read and react to human social cues. For example, I have seen in my classroom occasions when a student may be getting too close to the dog's face and the dog may start to move around a little bit. The student recognizes the dog is uncomfortable and they adjust their behavior accordingly. The largest success I have seen from the program is how the therapy dogs boost my students' self-esteem. Building students' confidence and self-esteem has contributed to their overall reading success by making them feel capable and competent.

In conclusion, the integration of therapy dogs into our school has been a remarkable and impactful tool for enhancing students' reading proficiency and SEL growth. By offering emotional comfort, fostering positive interactions, and nurturing a culture of empathy and understanding, therapy dogs have significantly contributed to the overall success and well-being of our students.

Book Corner

A Stone Sat Still by Brendan Wenzel

This is a delightful book for a read aloud, encouraging students to appreciate a simple stone from many perspectives. This book could work with many students and a variety of levels. The message of the book is one of perspective and hope. As time moves forward, the stone becomes many things to many elements of the environment. It can be big or small, rough or smooth, wild or a home (Wenzel, 2019), depending on whose perspective it is from. This opens the door to many conversations. Stones are common outdoors and often overlooked. But to some organism (an ant perhaps), this is a home. Questions can be posed to students, such as: When have you had a time in your life where you felt overlooked, but then someone saw a special quality in you to make you feel valued? Do you do this for others? Do you do this in nature? Another activity includes having students find a stone and describe it according to what they value: Is it smooth? Shiny? Flat, to cover the palm of their hand? Why do they think this is meaningful to them? What is a home? If this activity is followed with a mindfulness reflection, it might sound like this:

- Take a deep breath and begin to clear your mind. Close your eyes, or find something to focus on that will not let you be distracted by noises around you.
- Feel the stone you've chosen in your hand. Choose a quality that you appreciate about your stone. Let that quality settle into your mind and think about why it's important to you.
- Choose an element of nature that you are experiencing right now. It could be the breeze, the sun, the feeling of grass in your fingers . . . and then imagine your chosen stone in that element. Let your mind fill with gratitude for the beauty of nature. Take four cleansing breaths and open your eyes.

Hurry Up! A Book about Slowing Down by Kate Dopirak and Christopher Silas Neal

This is a wonderful book that highlights, from a child's perspective, the push and pull of living a busy life at home and school. This is a great book to read outside and have students share feelings they have when they have to hurry up. The second half of the book gives ideas about how to slow down and be mindful, most of which are outside. For example, one illustration is of a child gazing up at the sky. This is a wonderful prompt for a class activity of observing the clouds and asking what they see. Another page has them "make a wish," "take a walk," and "listen to the forest talk." These are opportunities to encourage social-emotional learning, with questions such as, "If you were to wish something for nature what would it be?" or "If you were to wish something for somebody else, what would that be?" This book does not have a lot of text, so it is easy to integrate small mindful activities within the reading of it.

More Children's Book Recommendations

- *What Does It Mean to Be Present?* by Rana DiOrio
- *Listening with My Heart* by Gabi Garcia
- *Rachel's Day in the Garden: A Kids Yoga Spring Colors Book (Kids Yoga Stories)* by Giselle Shardlow
- *Mindful Me: Mindfulness and Meditation for Kids* by Whitney Stewart
- *I Am Peace* by Susan Verde
- *Slow Down and Be Here Now: More Nature Stories to Make You Stop, Look, and Be Amazed by the Tiniest Things* by Laura Brand and Freya Hartas (illustrator)
- *The Listening Walk* by Paul Showers
- *You Belong Here* by M. H. Clark

CHAPTER 2

Nature and Language Arts
A SYMBIOTIC RELATIONSHIP

> *In my mind . . .*
> *Turning down the car sounds*
> *Turning up the bird sounds.*
>
> —Luca

The integration of nature into the language arts curriculum offers many benefits for educators and students alike. Students can explore the beauty and complexity of nature while honing their creativity and language arts skills. Furthermore, by embracing nature-based learning, we empower students to become environmental stewards, supporting a sense of responsibility for the natural world. This chapter outlines just some of the ways in which nature-based learning can be integrated into the language arts curriculum, including descriptive writing, storytelling, research, and expository writing. We highlight a poetry unit Helen created with a poet-in-residence, Colby Cedar Smith.

The Joy of Children's Books

Children's books are a powerful vehicle for exploring nature and fostering an understanding of our relationship with nature. Reading and analyzing works of literature with nature themes can deepen students' appreciation for the natural world while developing critical thinking and literary analysis skills. For example, folklore and mythology introduce them to the cultural dimensions of nature, helping them appreciate the diverse ways different societies have revered and interpreted the natural world. We included many of our favorite books in this chapter and throughout this book, but there are so many wonderful nonfiction and fiction books available. For example, the Blueberry Awards book lists are a great place to find great environmental children's books!

Nature-Based Writing Prompts

Nature provides sensory experiences that can inspire descriptive writing. Students can explore the sights, sounds, smells, and textures of nature and use vivid language to paint a picture for readers. Descriptive writing about nature can also help develop observation skills and enhance vocabulary. Creating choice in writing prompts is an important differentiation strategy that provides students of all abilities to work at their own pace. Table 2.1 shows some examples of prompts that can help students focus on the outdoors. The prompts are open so students can either answer in a straightforward way or they can go deep and metaphorical. They are also adaptable to the time of year. They are recommended for grades three to five but, depending on the group, could also work in second grade with some extra scaffolding.

Table 2.1. Descriptive Writing Prompts

Describe a **sweet** noise that you hear. (Usually we use sweet with our sense of smell, so you might need to stretch your brain for this one!)	Pick a leaf, a flower, a blade of grass, or a handful of dirt. Write what it **smells** like. Use similes to describe the smell.
Collect a series of **round** objects. Write how they connect. Are they all brown and smooth? Do they remind you of animal eyes?	Describe a **system** in nature. Write about the interconnectedness that you observe but use descriptive adjectives to make your writing sound artistic rather than scientific.
Choose a **pattern**. It could be spiral; it could be light, dark, light, dark. Identify places that you see this pattern in nature.	Create a **riddle**! Pair up with a partner. Each person finds an object for the other person to hold without showing it to them. Without looking, have one person describe the object while the other writes it down. Can you turn it into a riddle?
Look for items that are **yellow.** Describe the items without telling your reader what the item is.	Write a **pledge** to something. It could be a pledge to the sun, the soil, microbes, or the trees.

Storytelling and Nature Narratives

Students can create imaginative narratives set in natural landscapes, incorporating elements of adventure, discovery, and environmental themes. Writing nature narratives allows students to develop storytelling skills and explore their connection with nature.

Nature Sounds: Have children create and narrate their own stories inspired by nature sounds. Encourage them to incorporate phonetic words and sounds into their storytelling. They can imitate animal sounds, describe the sounds of different natural elements, or use descriptive language related to the phonetic sounds they hear in nature.

Fables: Since a fable is a story with animal characters that uses a moral to teach a lesson, a great place to find inspiration for this is outside. Read a few inside (or even outside) and then take a walk to look for inspiration. Assign an at-home walk at night with adult supervision to check out the nocturnal animal scene!

Animate Insects: Insects are not hard to find outside. With a few trowels and bug boxes in hand, all kinds of critters can be found. Why not do some anthropomorphism in your classroom and assign human attributes to the insects? This could be a great lesson on point of view as well!

Picture Prompts: Photography and art books such as *Andy Goldsworthy: A Collaboration with Nature* can provide imaginative prompts for fiction and nonfiction writing, as well as inspiration for creating nature-based art.

Phonetics

By combining phonetics instruction with nature-based learning, children in the younger grades can develop phonemic awareness and expand their vocabulary. Here are just a few ideas:

Phonics Practice: Bring your phonics warmup outside and conduct the same lesson you would inside or, even better, do a double-dip of phonetic practice and teach the lesson inside, and then have a digraph hunt outside where the students find the nature match to the essential skill that you that you taught earlier in the day!

Nature Sound Hunt: Take students on a nature walk and encourage them to listen attentively to the sounds they hear in the environment. Guide them to identify different natural sounds. Afterwards, engage in a group discussion where children can practice identifying and imitating the sounds they heard.

Nature Alphabet Hunt: Have children search for objects or features in nature that represent different phonetic sounds or letters of the alphabet. For example, they might find a smooth stone for the S sound or a lovely leaf for the letter L. Encourage them to name the object and its corresponding sound or letter as they discover them. They can also practice making letters out of sticks or other natural objects.

Letters with Leaves: Use chalk and draw on sidewalks and have students line the leaves up to create the letter or number of the week. Working on diagraphs? Make some leaf blends! When this multisensory activity has run its course, have them pick up the leaves and have a leaf dance party to celebrate their success. Leaf confetti is the best! If you want to bring nature inside after this activity you could read *Leaf Man* by Lois Ehlert and use the leaves to make a person, animal, or image to bring home.

Research and Expository Writing

Nature offers a wealth of topics for research and expository writing for upper elementary students. Students can explore environmental issues, scientific phenomena, or historical perspectives related to nature. Researching topics using nonfiction texts and expository writing about nature helps develop research skills, critical thinking, and the ability to effectively communicate complex concepts.

Students at HES connect writing with advocacy skills in a fifth-grade interdisciplinary project aligned to NGSS 5-ESS3-1 (*Obtain and combine information about ways individual communities use science ideas to protect the Earth's resources and environment*) and numerous Common Core ELA Standards, such as Reading Informational Text (RI.5.1), Integration of Knowledge and Ideas (RI.5.7 and RI.5.9), and writing standards such as W.5.8 and W.5.9, which focus on gathering evidence from sources.

Students begin by investigating and using inquiry-based research skills outside on school property. Typically, we begin with an indoor brainstorm; however, it can be powerful to do it outside, brainstorming the potential issues that affect their school and community. They might see air pollution from trucks in urban areas; they might notice trash or litter on the school grounds; or perhaps they notice how humans are *positively* impacting the environment. By leaving this initial research-based activity open for inquiry, often what the writing students create is more meaningful, more relevant to their lives, and they are more interested in writing about it. For this activity, teachers can offer students a variety of scaffolded tools, such as a concept map with the words "How humans impact the environment"—and then students complete the branches of the map.

After the concept map is complete, students narrow down their topic, research it, and create a website about the human impacts on ecosystems and the environment. The website connects each of their chosen issues under a larger umbrella. Students are tasked with informing their readers about their topic and the science behind it. They also provide a "call to action": a way that the public could help. The website is an exciting way for students to reach a broader audience.

Memorable Student Quotes from Nature Harmony Project

"I like the way this sand feels. I could write about that. I could use words. I could make a poem." Michael

"Should I answer this question as an adult or a child?" Jaylen

"This poem really defines me. This is my poem." Nate

"I had a dream, and my name was Lichen. I had another dream and I was a woodpecker. My head hurt the next day." Evan

"Could I find a poem in a dream? A lucid dream?" Luke

"For this poem, we're going back in time!" Conrad

"The birds are our assistant teachers today." Ana

"This is better than recess!" Christian

"I'm writing down a list of sparks in my notebook to inspire me later." Nathalie

"Play makes your mind open to new ideas." Jaylen

Nature Harmony Project: A Collaboration with Poet-in-Residence Colby Cedar Smith

Nature has long been a source of inspiration for poets. The beauty, cycles, and interconnectedness of the natural world lend themselves well to poetic expression. Students can engage in writing nature-inspired poetry, using metaphor, imagery, and symbolism to capture the essence of their experiences in nature. This poetry and nature unit was born out of a partnership between two people who love being outside and deeply enjoy nature as a restorative and deeply calming experience.

When Colby Cedar Smith and I (Helen) put our heads together, we looked at a need in the schools for kids to be able to be in the healing and restorative sunshine, fresh air, and beautiful weather. Colby, a professional poet and trained educator, said, "Wouldn't it be amazing to do an entire month of school where kids came to their language arts class, which was offered outside, and wrote poetry." We searched for grants and wrote an application to our state arts council for an artist-in-residence.

Frequently artists-in-residence bring dance programs, create murals, or develop plays. This unique idea was intriguing to the arts council and they funded our work. Poetry is part of the Common Core ELA standards at each grade level. Our unit focused on fourth grade; however, most of the lessons can be adapted to other grade levels. Although we taught it as a unit, we have presented activities here that can be stand-alone lessons. As part of the unit, we often connected the lessons to poems by established poets as well as children's books. Children observed the world around them, analyzed poems, and wrote their own poetry.

Photo 2.1. Poet Colby Cedar Smith (Helen Corveleyn)

I AM THE POET OF...

We began the unit by having everyone introduce themselves by choosing something in nature, adding a movement, and saying: "I am [name], the poet of . . ."

Examples of things the children (and teacher) chose:

I am Evan, the poet of chickens!
I am Caroline, poet of the spirit of the trees!
I am Mia, the poet of sunlight!
I am Nate, the poet of fuzzy pandas!
I am Mrs. Davis, the poet of birds!

CIRCLE POEM

> **The Circle**
>
> **by Colby Cedar Smith**
>
> *We come to the circle*
> *to open our minds.*
>
> *We let words*
> *flow like water.*
>
> *We plant them in the earth*
> *like wildflowers.*
>
> *We come to the circle*
> *to create*
> *and bloom.*
>
> *The circle of the sun.*
> *The moon.*
> *Our eyes wide with wonder.*
>
> *The circle holds*
> *our thoughts*
> *and emotions.*
>
> *It is strong*
> *and united*
> *like the circled arms*
> *of the forest grove.*
>
> *We come to the circle*
> *to center.*
>
> *Circle of the earth.*
> *Circle of ourselves.*
>
> *Circle of the tides,*
> *the seasons,*
> *our days together.*

Colby wrote "The Circle" for the residency and we began each class by reading it. The group would stand, sit, or gather in a circle, and we would spend the opening few minutes of our time together reading it out loud. This signified an acknowledgment of the norms we set in the beginning of the unit: a commitment to listening and writing with an open heart and open mind, a mutual respect for our natural setting, allowing ourselves to feel a range of emotions, and the recognition that the language that we use is descriptive and school-appropriate. We discussed the circle that we wanted to create together: a circle of trust and acceptance, and creativity.

For our first meeting together, we read the poem and then asked the kids to find circle-shaped objects in nature: dandelions, clouds, flower blossoms, wood rings, petals, clover leaves. After an invigorating hunt for circles, we crafted a collective poem. Students used their objects to create an individual verse of the poem on large chart paper. As each student contributed one line of poetry, a beautiful poem took shape on the page. Some lines were simple and some captured a more sophisticated tone. This collective poem helped to create trust in the teachers and the craft, because it took the pressure off the individual to create an entire poem from scratch and solidified our community of learners, emoters, and poets.

The reading of the poem set the scene for the remainder of the unit. The act of creating the collaborative poem melted away feelings of nervousness about writing poetry. The students instead felt a

> **The Circle**
>
> *Written by fourth-grade class*
>
> *The petal on a four-leaf clover*
> *A dandelion, holding the sun in my hand*
> *The end of a stick*
> *The cloudy sky*
> *A petal that looks like snow*
> *A leaf we find on the ground*
> *A turtle in its shell*
> *The jasmine that looks like a dancer*
> *A cut branch that symbolizes an end and a beginning*
> *The small particles of pollen*
> *A ladybug, and circles inside a circle*
> *A patch of moss*
> *A flower full of seeds, that will someday become another flower*
> *Pumpkins on a farm*
> *A tick*
> *A bumble bee*
> *An Earth full of nature.*

sense of accomplishment. In successive days, we read the circle poem in a variety of ways—all together as a group, whispered into a breeze, as an individual voice, everyone taking a line, two people alternating lines and, by the end, the kids could recite the poem from memory. It was one of their favorite parts of the unit.

BREATH

We began our lesson by focusing on our breathing during our mindfulness exercise. We plugged our ears and listened to our breath and heartbeat. We then read the circle poem and talked about stanza and how poets create space and breath in the spaces. We read the circle poem and did square breathing at each stanza space. Next, we ran wild across a field. We whooped and hollered, did cartwheels and somersaults. Then we plugged our ears again and listened to our breathing. Fast breath, panting, heart pounding. We concluded by listening to Franz Liszt's "Un Sospiro" ("A Sigh") and wrote what we were thinking and feeling.

TREES

Our tree lesson started with a discussion all about trees: their beauty, their age, a description of their textures, their utility in our environment, and their strength. Then the students explored trees—they touched, smelled, hugged, and, yes, even climbed on trees. We then read the mentor poem, "Whenever You See a Tree" by Padma Venkatraman, which is a concrete poem written in the shape of a tree. The

Photo 2.2. Student Poem: Trees and Leaves (Helen Corveleyn)

students worked on their own poems; we provided templates, or they could write in their own tree shape. This was a freeing experience for authors of all levels. Students of every ability were able to express themselves, and it led naturally to differentiation by giving a variety of templates or the choice of free writing, however they would like to design their poem.

POEMS FOR TWO VOICES

We began this lesson by performing "Water Striders" from Paul Fleischman's book *Joyful Noise: Poems for Two Voices*. In the book, each poem has two perspectives/voices. When you read the poem, it feels like a "call and answer" format, while sometimes the verses overlap and the voices say a line in unison. This can sometimes be challenging for kids but often ends in giggles of delight. To read the poem to the students, each teacher read a "voice." This is an important step that should not be skipped. Students followed along on paper, and benefited from hearing it performed out loud. It was an instant hit, and the students immediately wanted to try it themselves. We then had students dig in the garden or explore under a tree, searching for insects that inspired them. Explaining that humor is a great way to "celebrate nasty traits of insects" (think mosquitoes dressed up as tiny vampires "trick-or-treating" on human arms), the kids really get into this type of poetry. They beg not to stop—and all want to perform!

EVERYBODY NEEDS A . . .

This lesson started with a reading of the book *Everybody Needs a Rock* by Byrd Baylor and Peter Parnall, a delightful and funny poem about a child choosing the perfect rock. It is both instructional and humorous and written in ten steps, a very accessible format for students. As it is read, children always begin looking for a rock around them, which is a sure sign of students enjoying the poetry. We then explore our environment. We were at a local park for this lesson so students were able to search along the path, by the stream, in the gazebo, and by a reed pond. Students wrote list poems using the title "Everybody Needs a . . ." Some items chosen included: rocks, bees, giant reeds, mud, sticks, mulch, bark, flowers.

A possible extension to this whimsical poem could also be connecting the poem back to topics discussed in science. After they've written their first rough draft, incorporate a little science knowledge into their writing. Ask them to slip in a line about pollination if they are writing about flowers or pollinators! Tell them that a decomposer or two might be a fun addition to "Everybody Needs Dirt."

Everybody Needs a Bee

by Carys, fourth grade

Everybody needs a bee.
A bee buddy.
Because bees get lonely.
Which is why *YOU* should be friends with a bee.
Everybody is afraid of bees but they don't know why.
Maybe it's because yellow and black is out of fashion.
But bees want *you* to be their buddy.

(continued)

> Here's nine ways to make sure you select the correct bee buddy:
>
> 1. Make sure you know if you want to have a bee that can play with a certain *stinger* or one that can just eat, and fly.
> 2. Bees can't play dress up, so don't try. They don't make clothes bee sized.
> 3. It takes a while to find a bee buddy, so don't rush and take your time.
> 4. Remember that bees have issues if they're not in a circus with a dandelion, so give them a little tent and a hoop.
> 5. Bees like sweets! Give them plenty of fresh honey!
> 6. Your bee might be royalty, make sure they have a feather pillow and that they have a tissue box for if they sneeze.
> 7. Bees can be a little self-centered, so read to them a book about themselves!
> 8. Bees can be squashed. Or smashed. So be careful and don't ever use a fly trap!
> 9. All in all, bees are one of the best buddies.

A STUDY IN PERSPECTIVE

> *Think of a place in the natural world where you feel safe. You are walking through the forest, on the beach, or you dive into the ocean. You find a space where you feel safe. A creature is walking toward you, someone you really trust and you want in the safe space. You feel very safe and loved. Think of how the creature that you are imagining fits into your happy and comfortable space. Stay there for a little while to enjoy the feeling you get from that thought.*

This guided meditation provided a calm and imaginative start to our lesson. The students shared their experiences of their creature in the safe space, for example: "I was on the beach listening to the waves with my mom" and "I was on the deck of a cruise ship and a turquoise woman/dolphin spirit rose up from the water." We then read the book *They All Saw a Cat* by Brendan Wenzel and talked about how every character saw a cat, but they saw a different cat. The students, on their own, delved into a discussion about perspective and biases. We talked about how we all see things differently. We then wrote poems from an animal's perspective. We asked the students to think about living in an animal's body: What does the animal want? What does the animal need? What emotions does the animal feel? Then we asked them to write a poem about this animal encountering a problem. The kids wrote poems about: a panda being stolen from the forest and put in a zoo; a dog encountering a vacuum cleaner; a turtle trying to cross the road amid traffic; a kitten who gets adopted and taken from his mother.

We differentiated this lesson because the prompt seemed too abstract for some of the students. Students that had a hard time imagining themselves as an animal benefited from making the exercise a guessing game. The children wrote poems that describe an animal and everyone else got to guess the animal. We modeled it for the children using this example: "I have four paws that gently creep on top of the kitchen cabinet so that I can pounce joyfully on my prey." This helped to make the poem more concrete. They were SO excited!

MEANDERING

We began by telling the students that we are taking a "meander, not a hike, and we're going to stop to notice things as we go. We meandered, observing and listening to the world around us. At one point, we asked the kids to put words to what they imagine the birds are saying. The kids heard, "Whisper-will, whisper-will," "Suspicious, Suspicious!" "ACK! ACK! ATTACK! ATTACK!" and "How are you today?" in the bird calls. We then had students share what they were noticing and describe it in words. I (Helen) shared too!

> *"The shadow of the leaves looks like lace."*—Miles
>
> *"The buttercups sing their operas."*—Helen
>
> *"The birds are singing the first stanza of the poem!"*—Dominick

We (Colby and Helen) then did a dramatic reading of "Instructions" by Neil Gaiman. This is a poem that is worth checking out. If you enjoy showing your dramatic side to your students, you will love this poem. As we read stanzas we walked along a forest trail. We linked lines of the poem to everything we saw on the trail: the old trees, the river, the bridge. We actually walked in a loop, and the last instruction of the poem is: "Return the same way you came." The students were both a bit freaked out that it worked out that way, and at the same time enchanted and entranced by the poem. They were so enthusiastic about this that they couldn't wait to write their own instructional poem!

WATER AND ENVIRONMENTAL STEWARDSHIP

> *"The snake isn't a snake, it's an oil pipe, but the snake acts like a snake in the book. It's both."*
>
> —Caroline

For this lesson, we started with a mindfulness exercise concentrating on water. We read the circle poem and talked about the parts of the poem that focus on water. We then discussed conservation and protecting creatures, objects, resources in nature "without a voice." Students are very adept learners when it comes to looking at a picture book and connecting to beautiful illustrations. *We Are Water Protectors* by Carole Lindstrom is a combination of exquisite illustrations by Michaela Goade and a cautionary indigenous message that is highly relatable to fourth- and fifth-grade students. When reading this book, we have powerful conversations about people's voices that are heard in our society and people's voices that are sometimes not heard. Students naturally connect insects, land, and plants with "unheard voices" and truly embrace the environmental empathy that, frankly, is sometimes hard to teach.

We wrote "five-line" poems about water (oceans, pools, lakes, waterfalls, bathtubs, drinking water, etc.). The poems were beautiful and insightful. Lindstrom's award-

> ### Family Support
>
> *You should hear [my daughter] talk about poetry. She absolutely loves it and could not stop talking about it. You have made a huge impact. I should have videotaped her. She walked in the door, and first thing she said was, "Mom, I just love poetry so much," then she pulled out her notebook and read me her poetry.*
>
> This was written by a parent. What a beautiful affirmation of a unit that was almost universally enjoyed. (Let's be real here—as hard as we try as educators, you can't please ABSOLUTELY everyone. . . . But this sure came close!)

winning book could be used to teach a lot of avenues for environmental stewardship, but one thing is for sure: it is not to be missed.

The Power of Poetry for Student Self-Expression and Well-Being

As the STEM teacher in my school, I (Helen) don't usually have deep emotional conversations after doing a design scenario for an engineering project. We may have conversations about failure, perseverance, or being a good team member, but rarely the emotionally charged conversations like those that come from teaching outdoor poetry. When Colby and I wrote the grant proposal for this work, we summarized its objectives:

> *To promote a sense of natural wonder and appreciation of the Earth in combination with improving mental health, the Nature Harmony Project seeks to continue and promote a culture of poetic expression. We seek to continue to explore the natural world through poetry and promote a deep self-worth through creative expression.*

After teaching the unit, we reflected that when you write a grant application that addresses mental health and self-worth, you'd better be well equipped to really address all of the facets of mental health that arise. For this reason, we connected with the guidance counselor and invited her to participate in our work, connect with students on their journey of self-discovery, and be there to support them if needs arose. Below, we have included just a few examples of the many students who benefited from this collaboration. Their names have been changed for confidentiality.

JP

JP pulled me over to his desk, his eyes filled with determination. "I don't want to type the tree poem—it didn't come from my heart," he confessed. "The other ones, the other poems I wrote with my whole self." He opened up about his deep, angry feelings, admitting, "I need poetry. I really need to express things I can't say to my parents. I feel hurt and angry all the time." He continued, with so much self-awareness, "I feel like poetry is really good for me. I'm turning over a new leaf. Poetry is helping." With the support of the school counselor, JP's poems became a cathartic outlet for his emotions, helping him heal and grow. His journey taught us the importance of creative expression and a caring environment for young minds.

WALTER

We had a student who always wanted to concentrate on poems about darkness. His final poem was called "Walter's Death Poem," and had visions of hell, demons, and imagery of evil. We allowed him to express this within the confines of what is appropriate for school without limiting expression or preventing his emotional outpouring. Throughout the unit we kept a close eye on his emotional regulation and the struggles he had with expressing his feelings. With the guidance counselor on hand, we watched how Walter grew and worked through his feelings through the written word. Near the end of the unit, he finally volunteered to read his work to the class. The poem had "stormy" undertones, but the class was supportive and complimentary about his use of figurative language and some of the poetic decisions he made. This feedback created a shift in Walter. Acceptance and compliments from his peers changed his body language and you could see in his face the brightness of this acceptance. He quietly approached us after sharing his poem and told us he had more in his mind that needed to come out. He went to the corner of the outdoor tent to write a new poem for the next ten minutes. Ignoring everything else that was happening around him, he created the following poem:

> *Light can be a spark.*
> *Something that suddenly hits you.*
> *But you don't always realize it.*
> *But sometimes you do realize it.*
> *If you realize it . . .*
> *Take it.*
> *Hold it.*
> *And use it.*

GUS

Gus is a neurodivergent learner who sometimes struggles with emotional regulation. When we wrote the "poems for two voices," he was really focused on wanting to write a poem for three voices. But he couldn't find friends who also wanted to write using three voices. He didn't want to join another group. He felt really frustrated. He walked around the circle saying, "I don't WANT TO DO THIS." At the end of class, he came up to me and said, "I'm sorry I couldn't write today. I will write next time." The next session, when we were listening to music and writing, he came in late from another therapy. He felt disoriented and said, "There's nothing to write. I don't want to write! This is silly!" We asked, "Do you see any images while you're listening?" He answered fervently: "No! I feel angry!" To which I replied, "Great! Write about your anger!" He paused, looked up, and asked, "Really?" And I said, "YES—write it down!" And he wrote a really angry and wonderful poem:

> *I am bored bored bored.*
> *I am tired tired tired.*
> *I hate school school school.*
> *I want to go home home home.*
> *I want to play video games.*
> *Ok?*
> *Ok?*
> *Ok?*

Everyone cheered when it was read out loud to the rest of the class! This was an unexpected reaction for him. He quietly looked up and said, "Wait?? That was poetry?!" After that day, he still struggled—the experience certainly didn't solve all of his aggravation—but he did have an increased desire to try.

Book Recommendations

- *Call Me Athena* by Colby Cedar Smith
- *The Book of Nature: Wordsworth's Poetry on Nature* by William Wordsworth
- *Devotions: The Selected Poems of Mary Oliver*
- *Peace of Wild Things* by Wendell Berry
- *Andy Goldsworthy: A Collaboration with Nature*
- *Sing a Song of Seasons: A Nature Poem for Each Day of the Year* by Fiona Waters and Frann Preston-Gannon
- *National Geographic Book of Nature Poetry: More than 200 Poems with Photographs That Float, Zoom, and Bloom!*
- *The Story Orchestra: Four Seasons in One Day—Press the Note to Hear Vivaldi's Music (Volume 1)* by Jessica Courtney-Tickle
- *We Are the Water Protectors* by Carole Lindstrom and Michaela Goade
- *Listen to the Language of the Trees: A Story of How Forests Communicate Underground* by Tera Kelley and Marie Hermansson
- *Everybody Needs a Rock* by Byrd Baylor
- *They All Saw a Cat* by Brendan Wendzel
- *Joyful Noise: Poems for Two Voices (a book of poetry for children)* by Paul Fleischman.
- *Odder* by Katherine Applegate, illustrated by Charles Santoso
- *Grandfather Twilight* by Barbara Helen Berger

CHAPTER 3

Schoolyard Dependables
ESSENTIALS FOR OUTDOOR SUCCESS

Instructions for living a life.
Pay attention.
Be astonished.
Tell about it.

—Mary Oliver

This chapter focuses on nature that can be found in all schoolyard environments, whether blacktop, fields of green, snow, or sandy soil. We include lesson ideas and activities that encourage students to observe and investigate the fascinating world that is all around us in all kinds of weather.

Wonder-Full Investigations

Just about anything can be a source of study. For example, on the HES playground there were several holes in the ground. The kindergarten students wondered about the cause. This served as an exciting mystery. Every recess period, a large group of children could be seen discussing, puzzling, and pointing to the holes. We created a formal science lesson based on their interest. Students generated questions and hypotheses about the cause of the holes, and we gave them resources and tools, including magnifying glasses, scat and animal-track field guides, rulers, and tape measures. We even (carefully and without touching) analyzed the animal scat found near the holes. Looking at undigested seeds, grass, and fur prompted authentic conversations about the food chain. To solve this particular mystery, we ultimately borrowed a wildlife camera and connected it to a nearby tree to take pictures throughout the day and night. Students loved checking the camera by day and finally capturing evidence of red fox activity at night. The wildlife camera was easy to use and provided amazing conversations that transferred into every area of learning.

> ### Things to Prepare before Venturing Outside
>
> It is essential to have a plan before conducting lessons outside. Here are some guidelines:
>
> 1. **Scope out the area** you will be using prior to bringing the students out: figure out where you are going and set up a perimeter for the students so they know how far they can go.
> 2. **Be prepared:** Good footwear, first aid kit, epi-pens, walkie talkie/cell phone with access to the main office.
> 3. **Let the administration know** where you are going and how long you will be gone from your classroom. Make sure you know the process: Do you need a permission slip, or is a mention in your classroom newsletter enough?
> 4. **Make an outdoor toolbox.** Include items such as clipboards, extra pencils, magnifying glasses, maybe even a disposable bag for picking up garbage along the way. Even if you are not planning on using specific tools, it's great to have a kit with you so that kids can see that this is a classroom activity and not a recess or recreational activity.
> 5. **Time.** This may seem obvious, but it's essential for a good lesson. Not being rushed seems like a luxury, but if you can borrow time from another lesson so that you have a bit of wiggle room, exploration is a bit easier to enjoy when you are not rushing around.
> 6. **Bring a whistle.** You (probably) don't often use a whistle in the classroom. This is a tried-and-true method for signaling for immediate student attention.

SPECIAL SPOT

Provide students with a napkin (large index card, construction paper—whatever you like) and instruct them to walk around until they find a place on the ground that will be their "spot." Have them open the napkin and place it on the ground and create a frame for it, using found materials (sticks, leaves, rocks, etc.). Students can then remove the napkin and closely examine the revealed space for a minute or two (with the option of timing). Following this observation, have them sketch or take a photo of the spot. This activity can be revisited multiple times during the school year to compare over time. (Thanks to Jeanne Muzi for this idea!)

NATURE SCAVENGER HUNT

There are many wonderful websites with free templates for scavenger hunts. Scavenger hunts can be seasonal or connect to content areas such as math and literacy. It is a great way to have students practice describing words and/or categorization. Always include a blank option on the scavenger hunt so the students can find something to share. In areas where it is best to leave nature untouched, students can have a clipboard with a checklist or guided worksheet for drawing, or take a picture of items. In places where they can collect, a small baggy or reusable container for collections is important, especially for younger students. Students love to create their own design with the objects they find.

> ### Teacher Books for Outdoor Activity Ideas
>
> - *The Nature Connection: An Outdoor Workbook for Kids, Families, and Classrooms* by Clare Walker Leslie
> - *Science Adventures: Nature Activities for Young Children* by Elizabeth Sherwood and Robert A. Williams
> - *The Book of Nature Connection: 70 Sensory Activities for All Ages* by Dr. Jacob Rodenburg
> - *Exploring Nature Activity Book for Kids: 50 Creative Projects to Spark Curiosity in the Outdoors (Exploring for Kids Activity Books and Journals)* by Kim Andrews
> - *Nature-Based Learning for Young Children: Anytime, Anywhere, on Any Budget* by Julie Powers and Sheila Williams Ridge
> - *Ultimate Explorer Field Guides* by National Geographic, a series of seven books on different topics including birds, insects, rocks, mammals, and trees.
> - *Nature Preschools and Forest Kindergartens: The Handbook for Outdoor Learning* by David Sobel

TREE RUBBINGS

Engage students in a hands-on exploration of trees and other surfaces through rubbings. Have students find a tree with interesting bark textures. They can place a sheet of paper against the tree trunk and use crayons or pencils to rub over the surface, capturing the unique patterns and textures. Students can explore multiple trees, comparing and contrasting the textures they discover.

THROW A NATURE PARTY

Instead of having a pajama day or funny-sock day, throw a nature party! Have the students create decorations with natural objects they collect. The party can be for the students and/or have a mini-party for animals or other creatures, such as fairies and gnomes. Students delight in finding tiny items; for example, in the springtime, fallen buds are a favorite, and in the fall, look for beautiful tiny leaves or pinecones. They can also make party hats out of sticks. Don't forget, all good parties have confetti! We give each student a handful of birdseed to throw under trees or in the grass next to where the students play outside or by a window with a view. Students love to watch the birds hunt for the seed. The best part of the party "decorations" are that they don't require cleaning up. Students can leave the decorations there for the next class or other creatures to enjoy.

INSECT APPRECIATION

Students love searching for insects. After observing the insect (and perhaps drawing it), students can identify one trait of the insect and share with the class why they admire the insect for that trait. Older students can use a dichotomous key to try to identify insects. These are easy to print for each region of the country, and if laminated they can be brought outside and used over and over again.

Praise BEES

It is valuable to promote insects that may scare children but are such necessary contributors to our environment. Students can create a "thank a bee" (or other stinging insect) card, poem, or sign for a writing prompt. A fun idea is to get some seed paper and create your message that you can then go and plant outside. They can also make signs to put in the garden and around the school grounds (or even a local park) that say things like, "Thank You for Pollinating!" or "Pollination Nation: Bees Welcome to Rock Out Here!" These projects develop students' appreciation for insects, while promoting their environmental stewardship.

Bring the Outdoors In: Making Terrariums

Terrariums are an exciting means of teaching students about habitats, aquatic and terrestrial ecosystems, and the water cycle. They are a great way to introduce and emphasize the importance of scientific models to study the natural environment. Mason jars make excellent terrariums, but other containers can be used as well—a spaghetti-sauce jar or a reused clear plastic container such as two-liter soda bottles. When using them in relation to teaching about ecosystems found in the schoolyard, teachers can have students collect natural items such as soil, sticks, flowers, leaves, and pebbles. Students can also have objects represent larger objects, for example sticks can be used to represent trees. Tools such as popsicle sticks or chopsticks are useful and fun for students to move around as they create their own mini-world. For older grades, terrariums are a powerful way to teach about the water cycle as the process can be viewed in a sealed jar.

Teacher in the Limelight: Keeko de la Pena

Keeko de la Pena is a fourth-grade science teacher at Terence C. Reilly School No. 7 in Elizabeth, New Jersey. She also serves as an adjunct professor at Ramapo College in New Jersey. This section is written by her and includes two of her favorite lessons.

Science is my professional passion! Traveling, spending time with my family, wildlife conservation, and horseback riding are my personal passions! Gardening lessons have been an important part of teaching the Next Generation Science Standards to all of my students and these are two of my favorites.

Photo 3.2. Teacher Keeko de la Pena with Beanstalks (Keeko de la Pena)

Mealworm and Butterfly Life Cycle

Mealworms undergo a complete metamorphosis, which is a rather interesting transformation. In my class, we examine the various stages of the mealworm life cycle. We learn about the basic needs of these insects and actually feed them lettuce and other vegetables from our garden. They teach us about the importance of recycling. We also take time to learn about their life cycle and compare their life cycle and structures to the butterflies'. The butterflies are released in our garden once they can fly.

Growing with Beans

This project requires the following materials:

- dry bush bean or lima bean seeds
- cups
- magnifying glasses
- soil
- Ziploc bag
- markers
- construction paper
- cotton
- ruler
- wooden dowel or chopstick
- Jack and the Beanstalk book

Beans are an ideal plant for students to learn about because they are easy to grow and care for. Growing beans in the classroom can be a fun and educational experience for students of all ages and I have adapted this lesson depending on the age group. The first lesson requires student observation of dry bean seeds before planting them in cups. Each student labels their cups and plants a seed. They create a castle and a cloud with construction paper and cotton balls. A Ziploc bag over the cup will help retain moisture and trap heat. The students can see whose beanstalk reaches their castle first.

Snow and Puddle Exploration

Tracks Scat: Students love venturing out into the fresh snow to see animal tracks. After a storm at HES, we had students bring in snow pants and boots and explore like "real scientists." It is amazing how much they learn from being with peers and doing something out of the ordinary in school. The tracks were an exciting mystery to solve, and the scat was like treasure. It was well worth the winter walk!

Soil and Mud Exploration: Encourage students to explore different types of soil and mud, discussing the varying textures, colors, and consistencies. Students can use natural materials like leaves, twigs, and flowers to create mudpies.

Build a Bridge: When a puddle forms on the school grounds after rain, challenge students to work collaboratively to design and build a small bridge using natural materials like sticks and rocks. This activity introduces basic engineering concepts and problem-solving skills.

Water-Drop Transfer: Provide students with pipettes and small bowls or containers. Encourage them to use the pipettes to transfer water from the puddles or snow into their containers. This activity enhances fine motor skills and helps children understand the basic principles of water transfer.

Ecosystem Exploration: After a heavy rainfall, take the opportunity to study the temporary ecosystems formed in puddles. Guide students to closely observe the puddles and the various living organisms that may have gathered there. Bring magnifying glasses, handheld microscopes, and empty ice cube trays to collect any specimens that might be hiding in the puddles. Remember to put them back quickly though; too long outside of their natural habitat may be a risk for them.

Children's Books about Trees, and Creatures Who Love Them

- *The Tree Lady: The True Story of How One Tree-Loving Woman Changed a City Forever* by H. Joseph Hopkins and Jill McElmurry
- *Because of an Acorn* by Lola M. Schaefer, Adam Schaefer, and Frann Preston-Gannon
- *The Branch* by Mireille Messier and Pierre Pratt
- *Tree: A Peek-Through Picture Book* by Britta Teckentrup
- *Can You Hear the Trees Talking? Discovering the Hidden Life of the Forest* by Peter Wohlleben
- *Do You Know Where the Animals Live? Discovering the Incredible Creatures All around Us* by Peter Wohlleben
- *Those Darn Squirrels!* by Adam Rubin and Daniel Salmieri
- *Stranger in the Woods: A Photographic Fantasy* by Carl R. Sams and Jean Stoick
- *Steve the Dung Beetle: On a Roll* by Susan R. Stoltz; illustrated by Melissa Bailey
- *The Leaf Thief* by Alice Hemming and Nicola Slater

Creatures of the Yard

SQUIRRELS

One of my (Louise) favorite books is Ben Mardell's *From Basketball to The Beatles: In Search of Compelling Early Childhood Curriculum*, written in 1990. I have used it in my courses with preservice teachers for the past twenty years. In the book, Mardell describes the exciting, meaningful units he taught with his preschoolers. Mardell uses story, play, and imagination to engage children in their learning. The first chapter is about squirrels—and it is very memorable. In addition to integrating music, story, art, and cooking acorns into their study of squirrels, he has a scientist come in to dissect a dead squirrel with the children. This is always met by astonishment and fosters great conversations with preservice teachers. The power of the chapter is in the way the children (and the reader) are awakened to the natural world in their schoolyard. They begin to pay attention to the squirrels and their behaviors. Mardell has students draw squirrels, tally the number of squirrels, and investigate the lives of squirrels. The topic takes on a life of its own as the children begin to spontaneously play, observe, and discuss squirrels during recess, free play, and at home. Nature provides ample opportunities for indirect and direct engagement with creatures in the immediate local environment.

Like Ben Mardell, Helen incorporates squirrels into her STEM teaching. Squirrels are highly observable and make for a great topic for science. Squirrels are also high-functioning assessors of risk in nature. A fantastic way to teach animal behavior (and meet the "Asking Scientific Questions" science practice of NGSS) is to have students of any age sit quietly and observe animals in nature. Squirrels will let us do this! I (Helen) tell students I am setting a three-minute timer. I tell them to sit as still as possible and write down everything they see the squirrel doing. Then I have them compare with a partner to see if they observed similar actions. This often leads to a discussion about cause and effect (a NGSS Crosscutting Concept). I ask students questions such as: "What do you notice about how the squirrels react?"; "What bothers the squirrels and what doesn't?"; "How do you know?"; "How do squirrels communicate?" Students are excited to learn and observe how squirrels communicate with their tails. I then have students brainstorm a list of questions. They tend to come up with interesting and high-level questions because they are invested in their subjects and they are *really* curious about squirrel behavior!

BIRDS STEAM PROJECT

At HES, we created a STEAM project to connect kids with backyard birds. Students learned about native birds and the importance of birds in our ecosystems. We looked at many of the songbirds that students see at bird feeders, at home, or in the schoolyard. We discussed three different important roles of birds in the ecosystem: (1) birds are pollinators, providing an important role as they disperse seeds; (2) birds are indicators of ecosystem health because of their ability to adapt and change according to

Photo 3.1. Nature Harmony Glowforge Wood Art (Helen Corveleyn)

environmental stressors; and (3) birds act as a "cleanup team" in nature, as scavengers for dead or decaying organisms. Students listened to a presentation about native birds in New Jersey and then created drawings of the birds that they found beautiful or that they saw regularly in their yard. Using those drawings, we uploaded them to the laser printer (a Glowforge, which is a great machine for art installations) and used different wood varieties to turn student art into a permanent installation.

Book Corner

There are numerous apps available (such as Merlin ID) that play bird calls and record and identify birds (such as Song Sleuth). However, there are also wonderful books that are child friendly and allow students to press a button to hear a bird sound, with information about each bird. For example, *The Little Book of Backyard Bird Songs* by Andrea Pinnington has twelve different common bird calls. The book is part of a series of books aimed at children that include books on woodland birds, wetland birds, garden birds, and "dawn's chorus." The Cornell Lab of Ornithology publishes a larger book suitable for older children and adults. These books can be used in so many ways. For example, they can guess what the bird may look like based on the sounds, and/or what they are communicating. They can listen to various calls by the same type of bird to discern how birds use sound to communicate different messages. They can draw visual representations of the bird calls. They can create a class field guide, and each focus on one bird to research. They can compare with bird sounds they hear outside and see if they can recognize the bird based on its calls. Children love to fly around and mimic the bird sounds! All of these activities develop student knowledge and observational skills in a fun and exciting way.

Recommended Children's Books about Birds and Nests

- *Bird Builds a Nest* by Martin Jenkins
- *How Birds Build Their Amazing Homes* by W. Wright Robinson
- *Even an Ostrich Needs a Nest: Where Birds Begin* by Irene Kelly
- *Urban Roosts: Where Birds Nest in the City* by Barbara Bash
- *The Call of the Osprey* by Dorothy Hinshaw Patent
- *Falcons Nest on Skyscrapers* by Priscilla Belz Jenkins
- *Nesting* by Henry Cole
- *Robyn Boid: Architect* by Maree Coote
- *Birds* by Kevin Henkes
- *Elliot and the Osprey Nest (Elliot's Adventures)* by Ingrid Simunic
- *Little Book of Wetland Bird Sounds* by Andrea Pinnington and Caz Buckingham
- *The Little Book of Woodland Bird Songs* by Andrea Pinnington and Caz Buckingham
- *The Little Book of Backyard Bird Songs* by Andrea Pinnington and Caz Buckingham
- *Bird Songs: 250 North American Birds in Song* by Les Beletsky

NESTS

There are so many ways to integrate nests into the curriculum at all grade levels. For example, making nests aligns with the NGSS for Kindergarten, K-ESS3-1: *Use a model to represent the relationship between the needs of different plants and animals (including humans) and the places they live.* Louise developed a lesson—"Where does the osprey live?"—as part of her work with Project Sea, a teacher professional-development grant that focused on marine science and climate change education. The focus of the osprey lesson is why and how birds build nests. It includes both the negative and positive aspects of human impact on osprey: the loss of osprey habitat and the successful effort by conservationists to build platforms for osprey nests. The central activity is having students build nests with natural materials.

There are many ways to introduce a lesson on nests, depending on grade level. For example:

- Go outside and search for nests. If students can't find any, that is a great point of discussion.
- Show examples of nests (if possible) and/or pictures of different kinds of nests, first in trees (see below for links), since that is their most likely observation, and then pictures of nests not in trees, and ask what they notice and why some birds may not have their nest in a tree.
- Ask students if they have ever seen a nest, who makes nests,, and why (shelter, eggs).
- Ask where birds put nests and discuss why they may choose the location (such as trees).
- Make a list of possible materials used for a nest. Explain that there are different ways of building nests and that many birds, such as osprey, build nests with large sticks at the bottom, using smaller materials toward the top, and that they line their nests with soft materials like moss and grass. This is a great opportunity to ask students why that may be the case.

When building nests, students should ideally collect materials outside, from the ground. The teacher can add sticks and other materials if needed. Students can either collect materials and create nests at the same time or first collect materials and build nests after all materials have been collected. If at the same time, the teacher can provide a flat box to groups of children or have them "fly" and bring natural materials to the nest; mud can be used to hold sticks together. They can take turns building and collecting. If collecting materials first, once they have materials collected, and a small bowl for mud, provide shallow boxes for them to build their nests. This can be done in small groups or pairs. To conclude the lesson, students can "fly" to each other's nests to share what they included and why.

Nest building can be adapted to make it an engineering design challenge. Some possible challenges include limiting time or materials, for example, having students use tongs to build the nest to represent bird claws, or challenging them to build a nest that can safely hold an object about the same size as a bird and their eggs. Other possible nest-related activities include:

- Have children make wings, beaks, and/or eggs in the art area to wear during activity and/or in dramatic play.
- Connect to conservation organizations and have a guest speaker come to talk about their work.
- Go on a "nest walk" and search for nests in trees (and other locations).
- Have children draw pictures of their designs.
- For younger grades, create a "nest area" in the classroom so children can pretend to be birds in the dramatic play area.
- Sing songs, rhymes about birds.

Resources for Osprey Lesson

Project Sea Website: https://sites.google.com/tcnj.edu/projectsea/home?authuser=0

Facts about osprey nests from Birdfact website: https://birdfact.com/articles/osprey-nesting #whatDoOspreyNestsLookLike

Article: "Why Birds Are the World's Best Engineers" https://www.nytimes.com/2020/03/17/science/why-birds-are-the-worlds-best-engineers.html

Article: "Some Welcome Summer Visitors: Ospreys Counts Are Up" by Judy Smestad-Nunn https://www.jerseyshoreonline.com/ocean-county/some-welcome-summer-visitors-ospreys-counts-are-up/

Pictures of different types of nests: https://www.researchgate.net/figure/Views-of-birds-nests_fig2_337317898

Resources from Conserve Wildlife New Jersey

- Osprey Project webpage: http://www.conservewildlifenj.org/protecting/projects/osprey/
- Osprey live cam: http://www.conservewildlifenj.org/education/ospreycam/
- Conserve Wildlife YouTube channel: https://www.youtube.com/user/ConserveWildlifeNJ
- Installing osprey platform article and video: http://www.conservewildlifenj.org/blog/2018/11/15/video-from-the-field-osprey-platform-install/

CHAPTER 4

Agriculture, Gardens, and Native Plants—OH MY!

In 2013 the National Center for Agricultural Literacy (Spielmaker & Leising) developed learning outcomes for agricultural literacy across the grade levels. They developed five themes with grade-level benchmarks that are aligned with national standards. We have included a table of their K–2 and 3–5 Themes and Outcomes (table 4.1). The benchmarks are organized by five themes that provide a useful framework for teachers to consider when planning their instruction.

- Agriculture and the Environment
- Plants and Animals for Food, Fiber, and Energy
- Food, Health, and Lifestyle
- Science, Technology, Engineering and Math Culture, Society
- Economy and Geography

The National Agriculture in the Classroom website (as well as many state affiliates) offers wonderful resources for teachers, including lessons, materials, and conferences. These lessons are aligned to standards and the benchmarks. Many of them also align to NGSS. There is so much we could have included within this chapter as each of the themes are a book in itself. We have chosen to focus on Agriculture and Environment Outcomes (Theme 1); Plants and Animals for Food, Fiber, and Energy (Theme 2); and Food, Health, and Lifestyle (Theme 3). In particular, this chapter looks at school gardens, fiber, and native plants.

Farming for Food: School Gardens

In 2018 Hopewell Elementary won the Best in New Jersey Farm to School Award. This award was due to initiatives such as the school's outdoor and indoor hydroponic garden, outstanding relationships with local farmers, and a homemade organic lunch program. School gardens have become increasingly popular in schools and are a wonderful way to engage students with nature and the food production process.

Table 4.1. National Agricultural Literacy Standards

	Early Elementary (K–Grade 2) Academic Content Areas: Social Studies, Science, Health	Upper Elementary (Grades 3–5) Academic Content Areas: Social Studies, Science, Health
Agriculture and the Environment Outcomes	T1.K–2 a. Describe how farmers/ranchers use land to grow crops and support livestock b. Describe the importance of soil and water in raising crops and livestock c. Identify natural resources d. Provide examples of how weather patterns affect plant and animal growth for food	T1.3–5 a. Describe similarities and differences between managed and natural systems (e.g., wild forest and tree plantation; natural lake/ocean, and fish farm) b. Explain how the interaction of the sun, soil, water, and weather in plant and animal growth impacts agricultural production c. Identify land and water conservation methods used in farming systems (wind barriers, conservation tillage, laser leveling, GPS planting, etc.) d. Identify the major ecosystems and agro-ecosystems in their community or region (e.g., hardwood forests, conifers, grasslands, deserts) with agro-ecosystems (e.g., grazing areas and crop-growing regions) e. Recognize the natural resources used in agricultural practices to produce food, feed, clothing, landscaping plants, and fuel (e.g., soil, water, air, plants, animals, and minerals)

	Early Elementary (K–Grade 2) Academic Content Areas: Social Studies, Science, Health	Upper Elementary (Grades 3–5) Academic Content Areas: Social Studies, Science, Health
Plants and Animals for Food, Fiber, and Energy Outcomes	T2.K–2 a. Explain how farmers/ranchers work with the life cycle of plants and animals (planting/breeding) to harvest a crop b. Identify animals involved in agricultural production and their uses (i.e., work, meat, dairy, eggs) c. Identify examples of feed/food products eaten by animals and people d. Identify food safety practices to demonstrate at home e. Identify the importance of natural resources (e.g., sun, soil, water, minerals) in farming f. Identify the types of plants and animals found on farms and compare with plants and animals found in wild landscapes	T2.3–5 a. Discuss similarities and differences in food, clothing, shelter, and fuel sources among world cultures b. Distinguish between renewable and nonrenewable resources used in the production of food, feed, fuel, fiber (fabric or clothing), and shelter c. Explain how the availability of soil nutrients affects plant growth and development d. Provide examples of specific ways farmers/ranchers meet the needs of animals e. Understand the concept of stewardship and identify ways farmers/ranchers care for soil, water, plants, and animals
Food, Health, and Lifestyle Outcomes	T3.K–2 a. Identify healthy food options b. Recognize that agriculture provides our most basic necessities: food, fiber (fabric or clothing), energy, and shelter c. Understand where different types of foods should be stored safely at home	T3.3–5 a. Describe the necessary food components of a healthy diet using the current dietary guidelines b. Diagram the path of production for a processed product, from farm to table c. Distinguish between processed and unprocessed food d. Explain the costs associated with producing and purchasing food e. Explain the practices of safe food handling, preparation, and storage f. Identify careers in food, nutrition, and health g. Identify food sources of required food nutrients

(continued)

Table 4.1. *Continued*

	Early Elementary (K–Grade 2) *Academic Content Areas:* *Social Studies, Science, Health*	*Upper Elementary (Grades 3–5)* *Academic Content Areas:* *Social Studies, Science, Health*
Science, Technology, Engineering, and Mathematics Outcomes	T4.K–2 a. Explain what tools and materials farmers/ranchers use to reduce heating and cooling in plant and livestock structures b. Recognize and identify examples of simple tools and machines used in agricultural settings (e.g., levers, screws, pulley, wedge, auger, grinder, gears, etc.)	T4.3–5 a. Compare simple tools to complex modern machines used in agricultural systems to improve efficiency and reduce labor b. Describe how technology helps farmers/ranchers increase their outputs (crop and livestock yields) with fewer inputs (less water, fertilizer, and land) while using the same amount of space c. Identify examples of how the knowledge of inherited traits is applied to farmed plants and animals in order to meet specific objectives (i.e., increased yields, better nutrition, etc.) d. Provide examples of science being applied in farming for food, clothing, and shelter products
Culture, Society, Economy, and Geography Outcomes	T5.K–2 a. Discuss what a farmer does. b. Explain why farming is important to communities c. Identify places and methods of exchange for agricultural products in the local area d. Identify plants and animals grown or raised locally that are used for food, clothing, shelter, and landscapes e. Identify the people and careers involved from production to consumption of agricultural products f. Trace the sources of agricultural products (plant or animal) used daily	T5.3–5 a. Describe how supply and demand impact the price of agricultural goods b. Discover that there are many jobs in agriculture c. Explain how agricultural events and inventions affect how Americans live today (e.g., Eli Whitney: cotton gin; Cyrus McCormick: reaper; Virtanen: silo; Pasteur: pasteurization; John Deere: moldboard plow) d. Explain the value of agriculture and how it is important in daily life e. Provide examples of agricultural products available but not produced in their local area and state f. Understand the agricultural history of an individual's specific community and/or state

Source: Adapted from Spielmaker, D. M., & Leising, J. G. (2013). *National agricultural literacy outcomes* CC BY NC SA (AP)

Simply stated, gardens can be a place of peace, hard work, and deep authentic learning. The benefits of gardening can be observed with any student who is brought into the garden to learn, to plant, to weed, or to harvest.

We have chosen to include just a few favorite lessons and garden activities for this section. Our goal is to provide ideas for both beginner and experienced gardeners. We also include ideas for teachers/schools where a large outdoor garden is not a viable option and/or teachers who want to expand beyond vegetable gardens.

Farm and Garden Resources

The best resource for learning about farms and gardens is people! If at all possible, a field trip to a local farm is a long-lasting, impactful experience for children. Having a farmer or gardener come to the school is always exciting. Most small-scale and homestead farmers are eager to come to the school for a day to share what they do and what they produce. In urban areas, there are often community farms and gardens.

There are many wonderful resources available for teachers online, such as the National Farm to School Network (https://www.farmtoschool.org/), which includes chapters for each state, as well as websites with school-garden resources, such as:

- The Edible School Garden https://edibleschoolyard.org/
- Life Labs https://lifelab.org/
- Wisconsin School Garden Network https://wischoolgardens.org/
- Kids Gardening https://kidsgardening.org/
- The Nature Conservancy https://www.nature.org/en-us/about-us/who-we-are/how-we-work/youth-engagement/nature-lab/school-garden-resources/
- UGA extension https://extension.uga.edu/programs-services/school-garden-resources.html
- Container Gardens https://www.thespruce.com/vegetable-container-gardening-for-beginners-848161
- What Are Fibers https://spinoffmagazine.com/what-are-natural-fibers/
- USDA https://www.fs.usda.gov/wildflowers/ethnobotany/fibers.shtml
- Woolmark Company's wool production and processing video https://www.youtube.com/watch?v=YwRbyTCqOQY
- Agriculture in the Classroom https://agclassroom.org/matrix/resource/402/

STARTING THE DAY IN THE GARDEN

Picture a cool late-summer morning in September. School is in full swing, summer is a distant memory. Students arrive at school with a variety of morning experiences. Some children had breakfast made for them, some have made breakfast for themselves and their siblings, some have not eaten breakfast at all. Some walk long distances or sit on a bus for an extended period of time. Some had a calm, happy morning, some had stressful mornings, while some children are experiencing trauma at home. Starting the day outside (even if it's only twice per week) gives all children something to look forward to and a strong start to their school day. Beginning the day with gardening allows the teacher to monitor the garden's needs (watering, weeding, nighttime critter activity) out of the heat of the later day. As a bonus, time outside allows students to jump, skip, or just get a few more wiggles out. It's a guarantee that their focus will be better, and the teachers get a few minutes to gather their thoughts, take a deep breath, and reset!

SUNFLOWER STRETCH

This activity is especially good for preschool to third grade. If there are sunflowers growing in the garden that works best, but other flowers will work as well.

1. Begin as a seed and have students squat on the ground and hold their knees. Say something like, "Life begins from the littlest seed. You are a tiny 'beginning' right now, waiting to wake up. Think of all the potential you have today!"
2. Sprinkle rain and add sunshine to your seeds. (A whimsical twist here for a surprise is to have a hidden spray bottle of water and mist your kids during this part. This is sure to gather screams of joy.) "Rain wakes the seed up! Begin to wake up and grow your roots! You might burst from your seed coat and begin to grow a stem to feel the warm sunshine surrounding you! You've decided to begin your life as a cheerful sprout!"
3. Students begin to stand and stretch their arms to the sky. "You are growing on the Earth! Your roots are getting deeper and you are growing a flower. Your flower searches for the sunlight and follows it in the sky. Stretch your leaves as tall as they will grow."
4. Continuing the stretch, "You are a full-grown flower, stretch your stem, your leaves, and your whole flowerhead to follow the sun from the morning, noon, and to the evening. Your stretch should let your whole plant grow!" You may choose to end the stretch here and begin your day stretched out and ready to learn.
5. You may choose to extend the sunflower stretch to include the entire life cycle of the flower by allowing birds to eat the seeds, bees to pollinate the flower, and insects to benefit from the stem.
6. As the sunflower drops its seeds at the end of its life, it returns to the original tiny seed ready to start a new beginning.

MORNING MEETING: GRATITUDE GREETING

Having morning meeting outside is a great way to begin the day with gratitude. When students greet each other, have them identify one thing in the garden that makes them appreciate the Earth.

FIVE SENSES WAKE-UP

Wake up your senses for a great day of learning. Be mindful in the garden using all five senses. Identify a student per sense and let them lead the meeting by offering an experience for that sense. Since you are in the garden, you can include taste.

Photo 4.1. Working in the HES Garden (Helen Corveleyn)

The Power of Physical Labor

It is invigorating to start the day with outdoor physical activity. Actually, any time outside is a break from the fluorescent lighting, stagnant air, and quite frankly all the smells that come with teaching children! When you feel your tolerance fading, and your students getting "itchy," it may be time for some good old-fashioned outdoor physical labor. Using their hands and working on a task provides a sense of accomplishment, belonging, and productivity at any time of the day. Activities can include raking, pulling weeds, turning the soil, or rotating the compost, to name a few.

I have never seen fifth graders so motivated as when I had five yards of mulch delivered to spread on an outdoor garden. They loved using the wheelbarrow to fill in the parking lot and bring it to the garden. They even figured out who would do which job—the fillers, the dumpers, the rakers, the directors, the traffic police. This was all voluntary during recess time! They begged me to get another shipment of mulch the following week! Who knew?

ANYTIME GARDEN ACTIVITIES

Nature Surprise: This is a wonderful activity found in Clare Walker Leslie's books. Identify a surprise from nature in the garden that makes you smile. Introduce it to another person and marvel at the beauty of nature.

Garden countdown 3-2-1 . . .: Pick three weeds, turn over the soil for two minutes, and give one compliment to a fellow gardener to make them smile and start the day on a great foot.

Would you rather . . . garden edition: Who doesn't appreciate a good game of "would you rather"? Each student comes up with a "would you rather" to pose to a partner. "Would you rather have a tomato plant with the sweetest tomatoes you've ever tasted, or a pumpkin grow so big you could win a contest?"

IT STARTS WITH SEEDS

Seeds are an incredible teaching tool. Portable, connective in science and nature, often inexpensive, abundant, reusable, truly an excellent manipulative and teaching tool. Here are just a few seed-based activities:

Seeds sounds activity (pre-K–2): This is a great activity for younger students. Providing a variety of seed envelopes is a fantastic way to "listen to seed sounds" and have students ask questions and make predictions. Try to vary the varieties, for example, sunflower seeds and nasturtium seeds sound very different from parsley and basil seeds. Cover over the picture of the plant on the front of the packet and label them with numbers. Then have students predict the size of the seeds based on their sound. Have them rank them in order, and then open them up to see if they could identify the larger or smaller seeds. This allows students to predict using a sense other than sight.

Spelling with seeds: Don't let sand spelling, shaving cream spelling, and rainbow writing have all the fun! Seeds are a great way to practice spelling or letter recognition. Give students a few cups of seeds and have them line the seeds up on the letters or spell sight words using the seeds. Great for fine-motor skills and for multisensory learning.

Data activities: Seeds are a great way to incorporate math skills such as data exploration with categorizing and sorting. Even in a mixed bag of birdseed there are several different types of seeds present. Sorting, categorizing, and lining seeds up from a heterogeneous mix is an introduction to graphing skills. When seeds are lined up next to each other, students can compare numbers more easily. For the upper grades, seeds can be used to collect authentic data. Instead of using the data set in a math book, hang a bird feeder in the garden. Use observations to count how many times birds visit one feeder versus another with different seeds in them. Use this to make a claim in science. (For example: "If I fill one feeder with just sunflower seeds, and the other with mixed birdseed, the one with mixed seed will have more visitors.") Then collect data on the feeders. This collection could be at different times of day, it could be done in shifts of students, it could be individual or as class data—but this invites real-world field research and may lead to more questions. Maybe birds or squirrels will not visit the bird feeder, preferring a dying sunflower. This in turn may bring up

AGRICULTURE, GARDENS, AND NATIVE PLANTS—OH MY!

questions about seed dispersal, animal behavior, or animal preferences. It can also lead to discussion of factors and variables, such as location and type of feeder.

Teacher in the Limelight: Naomi Montalvo

Our spotlight in this chapter is on Naomi Montalvo, a school garden coordinator and pre-K teacher at Juan Pablo Duarte–José Julián Martí School 28 in Elizabeth, New Jersey. She describes the joys and challenges of creating a school garden. She now has a flourishing garden and can be followed on Instagram @school28garden.

As a teacher in an urban school district, I worked with students to reflect on how we could maximize the limited space effectively to begin a school garden. Developing a school garden requires a great deal of planning beforehand. Our custodial staff played a crucial role in the initial planning stage. They assisted us in identifying utility lines to avoid, so the space would be safe. They also educated us on regulations we had to abide by, such as the no-dig policy requiring everything to be in a raised bed, container, or adding soil to create mounds.

As our planning process continued, we had to identify the water source and ensure that whatever we developed was within a reasonable distance. We observed which areas were shaded and which received eight or more hours of sun. Next, we had to think about the goals we wanted to achieve when implementing the garden. Green initiatives and the environment were very important to us, so we decided to make a pollinator

Photo 4.2. Learning in the Garden with Mrs. Montalvo (Naomi Montalvo)

garden. Having a limited budget, we started with a few barrel planters and repurposed old tires as planters. Members of our school community worked together to plant native plants and noninvasive plants to provide a source of nectar. Students also did research and planted host plants for local pollinators.

After much success, the second year we decided to add a fruit and vegetable garden. We fundraised and applied for grants to purchase raised beds, grow lights, and seed-starting materials. Students created fruit and vegetable wish lists, and together we decided what would actually grow in our hardiness zone. We created and followed a calendar of when to start each variety of seed and when to transplant them. After considering the space and light requirements for each plant, we implemented the "square-foot gardening" method to maximize our growing space and used a companion planting guide to know which plants to put together in a raised bed. Faculty members, students, and their families volunteered to care for the garden, even in the summer when school was finished. All fresh fruits and vegetables were donated to our families in need.

The following year we got help from a nonprofit environmental organization to learn how to make our garden more sustainable. With their help, we installed a drip irrigation system and used straw mulch to conserve water. We added several more native plants and planted milkweed for the monarch butterflies. We also installed a three-bin compost system to reduce vegetative waste and amend our soil. We try to repurpose things in our garden and limit the amount of plastic used.

Photo 4.3. Watering Plants (Naomi Montalvo)

Just as our garden was flourishing and students were working on a sensory garden, we were abruptly quarantined due to COVID and were not allowed in the garden for almost a year. During that time, our garden looked like a forest, overgrown with weeds, and a groundhog became a permanent resident. Unfortunately, we lost some plants and our wooden raised beds were in poor condition. Because the garden was inaccessible for so long, there was a great deal of interest in helping us get it back to what it once was. There was a big fundraising effort to replace the raised beds and buy materials to enhance the garden.

Since then, students have continued to develop the sensory garden, including adding edible flowers and plants with varying textures, scents, and colors. They also planted four dwarf apple trees in tall raised beds, creating a mini-orchard. They have added mounds of soil among the few trees we have to create a shade garden, incorporating plants that have low light requirements. And while it is in its beginning stage, they have created a cut-flower garden to create small flower arrangements for someone who may be in need of a kind gesture. None of the mentioned gardens take up an immense amount of space, as we want to demonstrate to our families that gardens can be created in even the smallest of spaces. Our school garden is truly special to us, not only providing our students with opportunities to learn about plants and interact with nature, but serving as a place to develop a sense of community, instill environmental awareness, lead healthier lifestyles, and improve mental health. And, for me, it is my home away from home.

Sensory Garden at Juan Pablo Duarte–José Julián Martí School 28

by Naomi Montalvo

Our school is located in Elizabeth, New Jersey, with a population of approximately a thousand students ranging from pre-kindergarten to eighth grade. Students worked together to extend the current vegetable and pollinator garden to include a sensory garden. This garden was created with the intent to give our school community the opportunity to engage with an outdoor space by using their five senses. Sensory gardens offer many benefits including: relieving stress, easing anxiety, fostering social interactions, sparking curiosity, promoting physical activity, strengthening fine-motor skills, and incorporating STEM activities outdoors.

There were several factors taken into consideration when planning. We had to ensure the plants we selected were nontoxic. We had to observe the duration and amount of sunlight the space was exposed to each day. We kept accessibility in mind to ensure everyone in the school had the opportunity to engage in the garden equally and added wheelchair-accessible raised beds and elevated planters. We also had to ensure that whatever additional accessories were used did not pose a hazard to the children, that is, small parts that can be swallowed and things that can break easily creating sharp edges.

Smell: We have included a variety of herbs such as: basil, oregano, curry, lemon thyme, rosemary, sage, lavender, chives, marigolds. We also have a dwarf lilac bush and fragrant peony in large pots.

(continued)

> **Touch:** We have included a variety of plants that have different textures such as: lamb's ear, yarrow, ornamental grass, celosia, hens and chicks, and juniper. These plants were selected to explore a variety of textures such as smooth, fuzzy, soft, rough, and spikey. A simple water feature was also added by using a large plastic planter with no drainage holes and a floating solar fountain.
> **Sound:** We have added plants and accessories that make sounds when the wind blows such as: a variety of wind chimes, a birdbath with a solar fountain, and grasses that make a swishy sound. Students also worked to create a sound wall with a combination of natural objects and repurposed objects.
> **Taste:** We have edible flowers such as nasturtiums, violas, pansies, calendula, borage, dandelions, and sunflowers. Some of the plants that offer edible leaves include mint, chives, parsley, basil, and spinach. The items that are allowed to be tasted are in a specified area of the garden with signs so that students know not to taste things elsewhere in the garden.
> **Sight:** We selected a variety of plants that have different color leaves and flowers including: portulaca, pansies, nasturtiums, sunflowers, petunias, marigolds, cosmos, snapdragons, miniature roses, dahlias, celosia, foxglove, and zinnias. We also have plants that come in a variety of shapes, heights, and growth patterns, such as plants that climb like climbing nasturtiums, plants that hang like *Tradescantia*, tall plants such as six- to eight-feet-tall sunflowers and short flowering plants such as dianthus. Students have also designed and painted several benches to add more color to the space.
>
> The final component is our sensory table located in the center of the garden. Middle school students have researched ideas and activities that we rotate biweekly to encourage students to explore natural materials including: rock stacking, nature rubbing, tree blocks, mud kitchen, leaf threading, weighing natural objects, art with natural objects, nature-themed tic-tac-toe, weaving station, leaf hunts, exploring gourds, and seed exploration.

Themed Garden Beds: Everyone Loves a Good Theme!

Gardens can be quite exciting when they are themed. It's fun for the gardener, visually appealing and tangible for the kids. Students can participate in all aspects of the garden, including the planning. Themes can be connected to the science curriculum, such as a pollinator or butterfly garden. Other ideas include sensory gardens, color gardens, tea gardens, salad gardens, and salsa gardens. Math can be easily integrated in planning the dimensions of the garden, as well as measuring and weighing growth.

Storybook Garden: In an exciting collaboration with the media center specialist, we had the idea at our school to make a storybook garden bed. She wanted to honor a classic, *Peter Rabbit*, but there are plenty of garden-themed books out there. In our storybook garden we planted lettuce, carrots, peas, and blueberries in the garden. When students found out that we would be sampling from the garden bed just like the classic Peter Rabbit, they took the storyline to the next level! Without direction or prompting, they hopped around like rabbits and began "stealing" from the garden. Truly, stories come to life for little minds when they are given the tools and setting for connecting to nature.

Flavor-riffic Garden: There are so many exciting flavors to explore with children in the garden. Lemon sorrel tastes just like Sour Patch Kids. Chocolate mint has the faintest hint of chocolate flavor and kids will spend ten minutes biting and smelling it just to know that they ate something green that tastes like chocolate. A natural sweetener, a Stevia leaf can be paired with lemon balm leaf to make "lemon drops." Students love to mix the herbs together in their mouths to create their own flavor combinations. They can even create their own advertisement for the flavor they created and "market it" to the class! This encourages public-speaking skills, advertising, and creative thinking.

Easy Peasy, Lemon Squeezey! Lemon balm, lemon basil, lemongrass, lemon mint, lemon thyme, lemon bee balm: plant an entire herb garden of just lemon herbs! The beauty of planting herbs is that they are low maintenance, last for a long time, and grow a lot of leaves small enough for little fingers. Bring in lemons (if you can't grow them in the garden area) and make lemonade or lemon tea as part of a math lesson on volume and fractions and head out to the garden to garnish with some fresh lemon herbs.

Pizza Garden: Plant basil, tomatoes, oregano, and parsley (sometimes even onion and garlic in half the bed, which is also exciting). Say to students, "This is the pizza garden" and they honestly look for pizzas growing out of the ground! This is always good for a giggle that leads to a robust chat about what makes a pizza taste good and the importance of fresh ingredients.

Community and Family Connections: Summer Garden Suggestions

Teachers with outdoor gardens often struggle with the question of what to do with the garden in the summer. Here are some suggestions!

Make your school garden a community co-op for the summer months. Have families sign up to farm the garden throughout the summer as their own community garden. Let them know that they can plant and harvest whatever they would like (or you could give suggestions based on your needs) while school is not in session. Provide access to the garden and create a rotating schedule for families to share the garden responsibilities if necessary. When school begins again, the families turn the garden over to the school and the teachers to resume the care, maintenance, and harvesting of the food. Host a garden "pot luck" at the end to thank all of the families who participated to enjoy a meal together. Make it fun—hang up some twinkly lights, or set a long table in the middle of your garden to celebrate a successful season. True community is best when it includes delicious food!

Create a summer community garden club. Pair with a local community group like Big Brothers/Big Sisters or a YMCA to use the garden as an addition to their camp program. Perhaps your school is the site for a town recreation camp. Make the garden a part of their curriculum!

Adopt a garden. Depending on the size of your garden, ask a family (or multiple families) to sign up for a week of gardening. During that week of the summer they are responsible for watering, weeding, and maintaining the garden. Establish harvesting parameters so that each week volunteers will be able to take a portion of the food home to cook. Uneven harvesting could lead to disappointment from volunteers, so be sure to establish ground rules in the beginning of the season for sustainable harvest sharing.

Native Plants, Meadows, and Pollinators

Often when thinking of school gardens and farming, vegetable gardens come to mind. However, flowers and native plants have huge potential for teaching and learning. This is also a great place to start for novice gardeners. Pollinator gardens create many opportunities for close observations and scientific discovery. A cutting garden allows for a unique opportunity for SEL connections.

THE GIVING GARDEN

The idea of a "Giving Garden" or "Gratitude Garden" is for children to grow and nurture flowers with the intent of giving a bouquet to someone to show gratitude. Many different ideas can "blossom" from this garden: upcycle vases from recycled water bottles or containers from classroom supplies that would otherwise be thrown out with notes of kindness and appreciation decoupaged onto them for support staff, custodians, or other educators in need of thanks. Have students find other leaves, branches, and grasses to add to the flowers to make a unique bouquet. They can even create small flags or "floral picks" with notes or whimsical art to add a special touch. One benefit is that these gardens extend the typical garden season and grow well in spring or fall with tubers, bulbs, or other types of perennial flowers. For example, experiment with pumpkins! Who doesn't love receiving a "pumpkin of thanks" with a tag attached saying, "*Gourd Vibes Only*" or "*Thanks for teaching the cutest pumpkins in the patch.*"

For kids, the act of giving is a powerful way to instill self-worth, a sense of purpose, and the joy of gifting something beautiful. This is a wonderful way to teach public-speaking skills, as the students can practice what they will say as they give their bouquet or nature surprise to the adult. As a teacher it is important to voice all of these aspects of showing appreciation to others, both as a group and individually with students. Asking questions such as, "How do you think Mr. G. will feel when he receives your special gift and hears what you have to say to him?" and "How will you feel when you give this to him?" It is a great writing lesson to attach letters to small bouquets of flowers for unexpecting recipients. Other recipients of letters and flowers include local small businesses, school administrators, PTA members, and guest speakers. The power of sincere thoughts written in unmistakable children's handwriting is precious!

CONTAINER GARDENS

Container gardens are a fantastic way to have an outdoor garden without taking on too much. It is also a good alternative if school facilities workers are weary of having a larger space that may look messy at certain times of the year. Any of the gardens described above can be planted in a container garden. For example, a container makes for a great tea garden, and a native-plant meadow in a container can act as a model of a larger ecosystem. Kidsgardening.com is a wonderful resource for lessons and guidelines for container gardens. The containers can be large or small, just about anything that can hold soil and have drainage holes drilled in the bottom will do!

Here are just a few ideas for planters. Many containers will work if there are holes drilled in the bottom for drainage.

AGRICULTURE, GARDENS, AND NATIVE PLANTS—OH MY!

- Plastic (or porcelain!) tub
- Half-barrel container
- Raised beds (these are also great as they are accessible)
- Wall gardens
- Stackable planter containers (there are many inexpensive options available)
- Repurposed items such as dresser drawers, baby pools, toy trucks

POLLINATOR GARDENS AND MEADOWS

Another wonderful way of introducing native plants into your school is creating a pollinator meadow. A garden takes patience and time to develop into something beautiful, but the results are worth it! When choosing the seeds or plant plugs to begin your meadow, keep in mind that you want a variety of native plants that will attract different types of pollinators. A great way to describe this to kids is something like this:

> *If your adult at home told you that they would only be making spaghetti and meatballs for dinner from now until you are a grownup, how would you feel? You probably would be ok with this for a little while (especially if that was your FAVORITE dinner) but eventually you would get tired of having the same old dinner night after night. You might suggest that you have a few different choices for dinner so that your body could get some different nutrients rather than just the nutrition that spaghetti and meatballs provide. Well, this is the same for birds and insects that pollinate this meadow. Some variety in the plants they have to graze on would be a welcome choice for these pollinators, and we have to do our best in this meadow to make sure we are offering them a variety of pollen and nectar from which to choose.*

The real beauty of planting native flowers for pollination is that they last for a long time, and that native plants are hearty. Even in extreme conditions, "native" really means best suited for the area. Native plants are also used to native conditions and can survive more swings in temperature, moisture, and soil quality. Returning to the Agricultural Literacy Outcomes, Theme 1, Agriculture and Environment Outcomes, focuses on the relationship of local ecosystems to agri-ecosystems as well as sustainable agriculture practices.

Community Connection: Pollinator Pockets Meadow Project

One difficult aspect of a pollinator meadow is waiting for a few years' worth of growth and knowing how to maintain the meadow. Frequently an area can be designated to turn a grassy area or a section of underutilized land into a meadow, but for the first few years it really looks like overgrown weeds. There may not be an abundance of flowers, and meadows are more active in the spring and summer than in the fall and winter. Something that can really promote success with a school meadow is collaborating with a local environmental group, the buildings and grounds crew at your school, and the team of students or teachers you bring with you along your journey. Often partnering with a school group is an attractive option for community conservation groups as it is hard to get youth involved.

In an effort to eradicate invasive species in a large meadow area, the local park association approached HES to see if students would be interested in helping to plant native plants in a large field of Canadian goldenrod. As an "Earth Month" event (because why not make Earth Day a month long), students from preschool to fifth grade were involved with planting small plots of native plants called "pockets." The intent was to plant natives to push other more invasive species out in a controlled environment. The park commission and a local conservation group planned the plantings. Students planted and the conservation association taught them how to massage the roots before planting and included a brief lesson, in child-friendly language, about why native plants are important to ecosystems. To engage students before the planting, students learned about the plant species and the pollination process along with the pollinators that would visit the plants. In art class, the students created a paper "pocket," chose the flower they wanted to create, and matched the pollinator.

Children's Books about Food and Fibers

- *Farmer Will Allen and the Growing Table* by Jacqueline Briggs Martin, Shabazz Larkin, and Will Allen
- *The Life Cycles of Butterflies: From Egg to Maturity, a Visual Guide to 23 Common Garden Butterflies* by Judy Burris and Wayne Richards
- *A Seed Is Sleepy* by Dianna Hutts Aston and Sylvia Long
- *Pumpkin Jack* by Will Hubbell
- *Miss Maple's Seeds* by Eliza Wheeler
- *The Flower Garden* by Eve Bunting
- *I Ate Sunshine for Breakfast* by Michael Holland
- *Plant the Tiny Seed* by Christie Matheson
- *Lola Plants a Garden* by Anna McQuinn
- *The Curious Garden* by Peter Brown
- *Up, Down, All Around* by Katherine Ayres
- *Plants Feed Me* by Lizzy Rockwell
- *The Tiny Seed* by Eric Carle
- *From Seed to Plant* by Gail Gibbons
- *Seed to Plant* by Kristin Baird Rattini
- *Growing Vegetable Soup* by Lois Ehlert
- *Eating the Alphabet* by Lois Ehlert
- *Planting a Rainbow* by Lois Ehlert
- *How Did That Get into My Lunchbox?* by Chris Butterworth
- *Harlem Grown: How One Big Idea Transformed a Neighborhood* by Tony Hillery
- *Extra Yarn* by Mac Barnett and Jon Klassen
- *Charlie Needs a Cloak* by Tomie dePaola
- *The Goat in the Rug* by Charles L. Blood, Martin Link, and Nancy Winslow Parker
- *Weaving the Rainbow* by George Ella Lyon and Stephanie Anderson
- *Sophie's Masterpiece* by Eileen Spinelli and Jane Dyer
- *The Weaver's Surprise* by Tom Knisely and Megan Lloyd-Thompson
- *Braiding Sweetgrass for Young Adults: Indigenous Wisdom, Scientific Knowledge, and the Teachings of Plants* by Robin Wall Kimmerer, Monique Gray Smith, and Nicole Neidhardt

Fabulous Fibers

Studying and discussing fiber is such an exciting way to connect children to agriculture and give them broader and deeper insight into the importance of agriculture historically and in the present. Fiber is truly interdisciplinary, connecting art, STEM, language arts, and social studies. It is a wonderful way to bring in ideas of sustainability as well. We have included just a few ideas for the classroom as well as resources. In particular, National Agriculture in the Classroom is a wonderful resource for lesson ideas and for ordering kits for teaching about fiber. Both the USDA and US Forest Service offer lots of great information for teachers and students on their website.

Finding Fibers: A great way to start is to have students think about fiber and its different sources. Simply having them look around their surroundings (such as rugs, clothing, etc.) and consider the source of natural and synthetic fibers can be an eye-opening experience. Using a magnifying glass or, if possible, a microscope, students can make observations on the fibers. They can also consider how they feel and smell, and make comparisons. They can categorize fibers based on attributes and discuss the different types of fibers: animal (wool, silk), plant (linen, cotton), and artificial. They can also bring samples of interesting fibers from home. Math can be incorporated by having students closely examine the different weaves, measure the length and width of different fibers, and for older students, calculate the surface area of woven fabrics.

Learning the Process from Farm/Nature to Store: A great hands-on activity is to have students explore samples of animal fibers (e.g., wool) and plant fibers (including stems, seeds, and bark), and watch a farmer or artist (in person or on video) process the materials. There are amazing videos available on YouTube, such as the Woolmark Company's wool production and processing video. There are countless videos on basket weaving, rug making, broom making, and quilting. Agriculture in the Classroom has numerous wonderful lessons available on their website.

Literacy Connections: There are numerous fiction and nonfiction books that incorporate fiber and fiber arts. Many favorite folktales and stories from all over the world include fiber, especially spinning and weaving, as part of the story. Fiber can be incorporated into the language arts curriculum in other ways as well. Students, especially in upper elementary, can research one item (baskets, clothes, rug, quilt, brooms, paper) and write the step-by-step process, or create a "how to" book based on the process. They can also interview a farmer, artisan, or even an older adult, such as a grandparent who works with fibers.

Activities with Fibers

Students learn best with hands-on activities, and with fibers there are so many to choose from! It is sad that too often "arts and crafts" and learning things such as "basket weaving" has been degraded so much in society that they are synonymous with frivolousness and are considered the opposite of "real learning." In fact, basket weaving and other crafts require a lot of dexterity as well as cognitive skills, particularly problem solving and mathematics. Here are just a few possibilities:

Cordage: Knotting, twisting, braiding yarn, thread (fine yarn) or rope (heavy yarn). Natural materials such as grasses and vines (grapevine, honeysuckle, wisteria) can also be used.

- *Social Studies:* Students can learn about the types of ropes and knots and their use, from shoelaces and friendship bracelets to boats and construction. They can try their hand at making simple knots (such as a square knot) and progress to more complex knots (bowline knot).
- *Math and Measurement:* Cordage is a great way to teach measurement concepts like length and circumference. Students can measure different objects around the classroom or outdoors. This can help them grasp the idea of units of measurement and compare sizes. They can also sort the pieces from shortest to longest or vice versa.
- *Science:* Cordage is perfect for studying tension and force. Students can pull on different types of cordage to see which ones stretch the most. Challenges include creating a pulley system using cordage and simple materials or building a bridge using only cordage and popsicle sticks. This can develop collaboration skills as students work together in groups.

Weaving and Knitting: Weaving can be a wonderful and creative activity for elementary school students. It not only enhances fine-motor skills but also allows students to explore various aspects of art, culture, and history (see chapter 9). Students can learn the different ways weaving is used by humans, including for functional and artistic purposes. Teachers (and students) can bring in items such as baskets, small rugs, and clothes to explore weaves.

Finger weaving or finger knitting is a traditional craft that is very popular in nature preschools and Waldorf schools. It is a calming activity and a great way for students to develop fine-motor skills and keep their hands busy. Weaving also provides an excellent opportunity to introduce mathematical concepts such as patterns and symmetry. Projects can be created individually or collaboratively, in pairs or with the entire class.

Flower Pounding and Pressed Flowers: Students can experience the beautiful colors derived from flowers through flower pounding or pressing flowers. Flower pounding has the added benefit of the enjoyment of making noise and using a hammer (wood, rubber, or other hammer). Simply put the flower and plants between two pieces of paper or cloth. Make sure to put paper towels or a cloth between the table and the paper or cloth so as not to damage the table. Hammer away! Flowers can be pressed under books or with a flower presser—although these require about two or three weeks to dry. Students can create a collage, greeting card, or bookmark with the dried flowers, or adhere the dried flowers to other natural objects such as sticks, rocks, clay, or cloth.

Students can experiment with flowers and natural dyes outside just by giving them a white piece of paper. If you ask them to create a message using only what is around them, they become incredibly creative using flowers to spell, sometimes writing with the flower petals. (Morning glories are quite "juicy" and act like a very satisfying purple paintbrush on white background paper!) Sometimes it just comes down to trying things to see if your kids will take it and run with it. Sometimes the temperature is just right—the kids need something novel to try and the creative juices are flowing. Often the best activities come out when you don't expect them!

Recommended Books

This is just a small sampling of the many wonderful books available related to farms, gardening, and native plants.

- *Farm to Table & Beyond* by Pamela A. Koch, EdD, RD
- *Growing Food* by Pamela A. Koch, Angela Calabrese Barton, and Isobel R. Contento
- *Gardening Projects for Kids: Fantastic Ideas for Making Things, Growing Plants and Flowers, and Attracting Wildlife to the Garden* by Jenny Hendy, Whitney Cohen, and John Fisher
- *Gardening for Children with Autism Spectrum Disorders and Special Educational Needs: Engaging with Nature to Combat Anxiety, Promote Sensory Integration and Build Social Skills* by Natasha Etherington and Jessica Kingsley
- *Growing Your Own Tea Garden: The Guide to Growing and Harvesting Flavorful Teas in Your Backyard* by Jodi Helmer
- *Learning Gardens and Sustainability Education: Bringing Life to Schools and Schools to Life* by Dilafruz Williams and Jonathan Brown
- *Wild Plant Culture: A Guide to Restoring Edible and Medicinal Native Plant Communities* by Jared Rosenbaum
- *Bringing Nature Home: How You Can Sustain Wildlife with Native Plants* by Douglas W. Tallamy and Rick Darke
- *Nature's Palette: A Color Reference System from the Natural World* by Patrick Baty, Elaine Charwat, Peter Davidson, André Karliczek, and Giulia Simonini
- *Wild Color, Revised and Updated Edition: The Complete Guide to Making and Using Natural Dye* by Jenny Dean and Karen Diadick Casselman

CHAPTER 5

Hydroponics and Aquaponics
CLEAN, GREEN GROWING MACHINES

Hydroponic and aquaponic systems are a wonderful way to bring nature into the classroom, especially for teachers who may be hesitant or have obstacles for outdoor gardening. As discussed in chapter 4, outdoor gardens have many academic, physical, and social-emotional benefits for students. However, outdoor school gardening programs can be hard to sustain due to factors such as the teacher's lack of time, funding, volunteers, space, and gardening experience, especially in urban environments (Ozer, 2007). Depending on the climate, it can be difficult for children to experience the growth of a plant from seed to harvest, particularly during the winter months.

Indoor hydroponic and aquaponic garden systems provide exciting educational opportunities in coordination with outdoor gardens (ideally), or as an alternative. Hydroponic and aquaponic systems grow plants without soil; hydroponics uses liquid nutrient solution while fish waste provides nutrients in the aquaponic system. Plants in indoor aquaponic and hydroponic systems have an accelerated growth rate and are not weather dependent so are easily integrated into schools year-round. There are a variety of options for growing indoors, from a simple mason jar to an extensive indoor farm setup. A school can set up a lab for students or purchase a mobile system that can be moved into multiple classrooms.

Hopewell Elementary School experimented with a flourishing indoor hydroponics system with weekly organic cafeteria meals featuring fresh vegetables from the hydroponic garden (year-round) and the soil garden (seasonally). There is also a thriving garden and environmental club. Helen's success at Hopewell has inspired many schools and teachers, including Louise, who brought hydroponics to her college for preservice teachers. Louise also developed a program that provides hydroponic kits and professional development to teachers in low-income urban schools. This chapter offers suggestions about how to get started and how to expand a garden, from tabletop to indoor farm. We also share ideas to support Social-Emotional Learning (SEL) competencies and learning in each of the content areas.

Planning and Setting Up

Like any new endeavor, in order to not get overwhelmed, start with a small setup. There are many systems to choose from; Amazon and other sellers have models with twelve plants for less than a hundred dollars. When possible, have the garden located in the classroom or a very accessible place in school. The larger models have wheels so they can be used in multiple classrooms. Make sure to consider the light and sounds of the gardens when setting up. Some of the larger models make noise and can be bright. The water reservoir of the garden requires refills so it is also important that the garden is kept near a water source. Smaller models can be filled with a pitcher and brought to a sink for cleaning. When selecting the growing medium, it is important to note that some rockwool brands may contain glass fibers that could pose risks to students. If possible, involve students in assembling the garden structure so that they can see all of its parts and understand how it works.

Suggested Materials List

- Hydroponics system
- Water pitcher
- Seeds
- Nutrients
- Rockwool or other growing medium
- Scissors/pruners
- Gloves
- pH kit
- Pollinators (made by students, or a makeup brush or toothbrush works well)
- Spray bottle
- Towels
- Hand vacuum/dustpan
- Mat for rug (for larger gardens)
- Stickers/paper/sticks for labels
- Plant log to organize and record growth
- Job charts for students

Types of Systems

There are many different types of hydroponic systems; we have included the most popular ones for schools.

MASON JAR/KRATKY METHOD

Mason jars are an easy way to introduce children to hydroponics and/or aquaponics. There are many resources available on mason jars and the Kratky method of growing.

Provide students with a thirty-two-ounce wide-mouth mason jar, a rockwool—or similar material—cube, a three-inch net pot, clay balls, and a short piece of string/cord (that acts as a wick from net pot to water; the cord is optional in hydroponics setup, but ensures that the rockwool will stay moist even if water is low). If focusing on hydroponics, nutrients will need to be added. Otherwise, a small fish such as a minnow or betta will provide the nutrients. Lettuce and herbs, such as basil or dill, will grow well. As an added activity if doing hydroponics, students can decorate the mason jar with burlap, paint, foil, or tissue paper, so that the roots are not overexposed to light.

DEEP-WATER CULTURE (DWC)/TABLETOP SYSTEMS

This is a great, low-cost way to bring hydroponics into the classroom. A small tabletop hydroponic system for twelve plants can be bought for about seventy dollars. These kits are usually DWC systems in which the roots are submerged in a reservoir. The system includes a pump to move the water around and LED lights. The kit usually includes the nutrients that need to be added periodically. It is also nice to have a pH kit to test the water, which costs about fifteen dollars. Students love to come in and check on their plants. Lettuce and most herbs grow very well in the system.

AEROPONIC GARDENS

In aeroponic systems, the roots are suspended in air with a pump that mists the roots with nutrient-rich water. Usually systems use vertical space to grow plants. Tower Gardens is a popular company used by schools. Stephen Ritz has numerous videos, a book, and other resources for using Tower Gardens in the classroom.

EBB AND FLOW AND NUTRIENT FILM TECHNIQUE (NFT)

These are systems that could be used in schools looking to grow a large crop. In the Nutrient Film Technique (NFT) system, plants are put into small nets or cups with a substrate to hold the seeds in place (Rockwool, oasis, or clay pebbles all work). The net cup sits in a trough or channel (there are a variety of options for this) and the nutrient solution is pumped over the roots from a reservoir. The channel is deep enough to allow oxygen to get to the roots. The water cycles through the whole system continuously for as long as you have the timer set for. Usually the nutrient solution cycles for fifteen minutes on and forty-five minutes off. Roots do not need to be continuously watered with this system.

The ebb and flow system is a bit more complicated, but not hard by any means. The benefit of this system is that you can change your plants' positions around pretty easily. This method includes putting the plant containers in a shallow tray (called a flood tray) where water is pumped in from a reservoir, "flooding" the roots of the plants and then emptying to allow for oxygenation. When using the hydroponics systems, it is always a perfect tie-in to talk about NGSS 5-LS1-1. (*Support an argument that plants get the materials they need for growth chiefly from air and water.*) Nothing teaches this in a more clear and visual way than hydroponic growing!

AQUAPONICS

In aquaponic systems, fish waste provides nutrients for the plants, and the plants help filter and clean the water for the fish. Aquaponic systems also provide a wonderful way for students to study ecosystems.

Maintenance

Creating a structured student job chart with tasks such as adding water, pruning, pollinating, adding nutrients, and testing pH (with gloves and supervision) is a great way to include students in maintenance while keeping track of it. This approach not only ensures the systematic care of the garden but also fosters a sense of responsibility and a growth mindset among students. Engaging students in problem solving when challenges arise further enhances their learning experience. While pests and diseases are less common in hydroponics, they do happen, so being prepared for such occurrences is essential. Information on combating common issues is available online. Joining an online hydroponic gardening community provides a valuable support network for troubleshooting and asking questions. It's important to work with facilities at your school to stay informed about power and electricity plans, especially during breaks when it can get dark and cold in the classroom. Planning the final harvest before breaks ensures a smooth transition and the continued success of the hydroponic garden.

What to Grow

Many herbs and vegetables thrive in hydroponic systems. Lettuce varieties such as kale, arugula, and romaine are well suited to hydroponics, offering a continuous harvest of fresh greens. Herbs and spices commonly used in cooking, including basil, rosemary, dill, cilantro, thyme, oregano, and parsley, all flourish in hydroponic systems. Additionally, herbs used for teas grow well, such as mint, lemon verbena, lemon balm, and lavender, although some may require a longer germination and flowering period. Peppers, cucumbers, strawberries, and baby tomatoes are also successfully grown in larger hydroponic systems. Furthermore, edible flowers such as nasturtium, borage, and calendula are beautiful—and fun to grow and eat!

Planning and Planting with Students

To minimize food waste and foster collaboration among students, it's useful to discuss plans for the harvest before initiating the planting process. Encourage students to envision how they will utilize their produce, whether it be for making pizzas, salads, smoothies, teas, or other culinary delights. Address the possibility of excess harvest and

Photo 5.1. Planting Seeds (Helen Corveleyn)

strategize on ways to utilize or share surplus produce within the community. For a more personalized connection to the garden, consider having each student plant their own seedling. Always record the plant's type and location immediately after planting. Ensure accurate labeling of each plant on a corresponding record chart, or if labeling on the system, use a pen or marker that remains legible even when exposed to moisture. Lastly, for optimal seed germination, place the seeds in darkness for twenty-four to forty-eight hours following the initial planting.

Curriculum Connections

SOCIAL-EMOTIONAL LEARNING (SEL)

Children love to check in on the plants each day and develop a close connection to the plants that grow. The gardens also support SEL. Students experience a sense of responsibility as they care and nurture the plants. They learn patience and experience the cycle of life. They develop a sense of community as they plant with their classmates and share the fruit of their labor in a salad-bration, or juicing party. Hydroponics and aquaponics also promote healthy eating and self-awareness of their nutritional needs as a living being. There are many opportunities for discussions about health and wellness, food access and local farming.

SCIENCE

One of the beauties of working with children is the joy of the unexpected. When we were first figuring out how to farm hydroponically with the fifth grade, I (Helen) had two vertical gardens in my classroom. Students took delight in watching the gardens bloom into food, and they were astonished at the fact that soil was not necessary for the plants to grow. Even though I explained it as we planted, they still had to see to believe. They made connections I never would have expected. This became apparent during a science lesson about ocean acidification and the increase in the ocean's pH caused by warming temperatures and increasing carbon dioxide in the atmosphere. One of the students from my hydroponic "water chemistry team" recognized the connection between the fluctuations of pH in the tower garden in relation to the growth of the plants and the acidification of the ocean water. This connection is not something that I would have expected from a fifth grader, but because they had authentic experience with hydroponics, they could apply their observations to other, more abstract scientific concepts.

Table 5.1 gives just a few ideas of the ways hydroponic gardens can be used to teach lessons that align to NGSS at every grade level.

Seed observation and analysis: Students can observe, compare, and make predictions about the seeds and their growth. Some seeds are extremely tiny while others are quite large. Students can also compare the germination of seeds in traditional soil and in hydroponic systems. They can record and draw their observations, discussing the differences and similarities in growth.

Plant parts exploration: Hydroponics give the opportunity to study the different parts of a plant and how they function. Students can use hand magnifiers and microscopes to study the plant parts. Flowering plants can be easily grown in the hydroponic system. (It is a fun experiment to not harvest lettuce or herbs such as basil and see the flowers!) Unlike soil gardens, hydroponics allows students to see immense root systems and it is so exciting for them!

Life cycle of plants: Since the life cycle of the plant is shorter with hydroponics, students can observe from seed to harvest of many plants and herbs in a month. Students can create visual timelines or even short animations depicting the different stages. They can experiment with extracting seeds from mature plants and planting the seeds.

Designing pollinators: Building a tool to pollinate indoor hydroponic plants is a great activity that aligns to NGSS 2-LS4-2. (*Develop a simple model that mimics the function of an animal dispersing seeds or pollinating plants.*) The lack of pollinators is a real problem that indoor hydroponic gardeners face, and devices are created to help

Table 5.1. Hydroponics at Every Grade Level

Grade	Next Generation Science Standard	Lesson/Activity
K	**K-LS1-1 From Molecules to Organisms: Structures and Processes** Use observations to describe patterns of what plants and animals (including humans) need to survive.	Students can observe the growth of plants and help document the maintenance of the plants in their science journals.
	K-ESS3-1 Earth and Human Activity Use a model to represent the relationship between the needs of different plants and animals (including humans) and the places they live.	By setting up a small hydroponic system, students can observe and help create a system where they add everything needed for a plant to grow.
	K-ESS3-3 Earth and Human Activity Communicate solutions that will reduce the impact of humans on the land, water, air, and/or other living things in the local environment.	When students see where food comes from, it promotes pro-environmental behavior. Local eating from farms or vertical farms reduces transportation pollution and is better for the climate.
1st grade	**LS1.A Structure and Function** All organisms have external parts. Plants have different parts (roots, stems, leaves, flowers, fruits) that help them survive and grow.	Watch the plant sprout from the seed: this is one of the best ways to see the cotyledon emerge from the seed coat. Also roots are highly visible through the whole process and look significantly different from when they are covered in soil.
2nd grade	**2-LS2-1 Ecosystems: Interactions, Energy, and Dynamics** Plan and conduct an investigation to determine if plants need sunlight and water to grow.	Growing plants in a soil-free environment. Observe and experiment with different light conditions.
	2-LS2-2 Ecosystems: Interactions, Energy, and Dynamics Develop a simple model that mimics the function of an animal in dispersing seeds or pollinating plants	Make hand pollinators out of pipe cleaners.
	2-LS4-1 Biological Evolution: Unity and Diversity Make observations of plants and animals to compare the diversity of life in different habitats.	Compare and contrast plants living hydroponically versus plants living outside. Study the fish in the aquaponic system.

(continued)

Table 5.1. Continued

Grade	Next Generation Science Standard	Lesson/Activity
3rd grade	**3-LS1-1 Heredity** Develop models to describe that organisms have unique and diverse life cycles but all have in common birth, growth, reproduction, and death. **3-LS3-2 Heredity** Use evidence to support the explanation that traits can be influenced by the environment.	Investigate hydroponic growing from seed to death. View the entire life cycle of the plant as well as the fish in the aquaponic system.
4th grade	**4-LS1-1 From Molecules to Organisms: Structures and Processes** Construct an argument that plants and animals have internal and external structures that function to support survival, growth, behavior, and reproduction.	Grow flowering plants like nasturtium—the flowers are easy to dissect and the eggs can be viewed under a microscope if you cut the ovule apart.
	4-ESS3-2 Earth and Human Activity Generate and compare multiple solutions to reduce the impacts of natural Earth processes on humans.	Hydroponics is a great way to talk about food deserts, food insecurity, and reducing water consumption for vegetation in a traditional farming setting.
5th grade	**5-ESS3-1 Earth and Human Activity** Obtain and combine information about ways individual communities use science ideas to protect the Earth's resources and environment.	Use hydroponics as an example of reducing the need for large land masses to grow vegetables; have discussions about water conservation (hydroponics use 90% less water than traditional farming).
	5-LS1-1 From Molecules to Organisms: Structures and Processes Support an argument that plants get the materials they need for growth chiefly from air and water.	Students see roots growing and plants thriving without soil; hydroponics is the most concrete demonstration to dispel student misconceptions about plants needing soil to live.

the process along. During the initial discussion, students should come up with the fact that bees are pollinators and that their bodies are a bit fuzzy. As they investigate, lessons can be layered into the engineering process about the structure and function of a bee or a moth's body being fuzzy to pick up the pollen, learning about the bee dance as they collect the pollen and considering ways that this process can be mimicked to create pollination indoors. Plants that grow berries work well for teaching pollination (strawberries, raspberries, or blueberries), although bushes can be difficult to grow indoors. Strawberries work in larger hydroponic systems and it's actually incredible to hand-pollinate strawberries and then see the shape of the berries when they grow; if you don't pollinate them the way bees do, you get silly misshapen berries in a "nature takes no prisoners situation"! Tomato plants, or any plant with pollen, also work. Pollinators can be made out of pipe cleaners, makeup brushes, or toothbrushes.

LANGUAGE ARTS

Descriptive language: Students can describe plants using sensory details to convey the sights, smells, and textures of the plants.

Plant journals: Students can create journals detailing the growth of a specific plant in the hydroponic system and/or documenting the care of the plants. They can describe the changes they observe and reflect on their experiences.

Children's Books

Up, Down, and Around by Katherine Ayres and Nadine Bernard Westcott
The Amazing Life Cycle of Plants by Kay Barnham
The Curious Garden by Peter Brown
The Tiny Seed by Eric Carle
Growing Vegetable Soup by Lois Ehlert
Eating the Alphabet: Fruits and Vegetables from A to Z by Lois Ehlert
From Seed to Plant by Gail Gibbons
Grow Happy by Jon Lasser, Sage Foster-Lasser, and Christopher Lyles
Lola Plants a Garden by Anna McQuinn and Rosalind Beardshaw
National Geographic Readers: Seed to Plant by Kristin Baird Rattini
Plants Feed Me by Lizzy Rockwell

Hydroponics vocabulary: There are many hydroponics-related vocabulary words for students to learn. They can label the parts of the system and create their own illustrated glossaries. Vocabulary may include the following words:

Hydroponics	pH level	Rockwool
Aquaponics	Reservoir	Oxygenation
Nutrient Solution	Transplanting	Germination
Roots	Photosynthesis	Hydroculture
Growing Medium	Soluble	Sustainable Agriculture
Aeroponics	Perlite	

Hydroponics in the Classroom Project

In 2019 I (Louise) was awarded an internal grant from my college, which provided the financial support needed to develop The College of New Jersey's Hydroponics in the Classroom project. The project focuses on the academic and social-emotional benefits of bringing indoor hydroponic gardening into early childhood and elementary classrooms. All of the schools are located in low-income urban communities. Through this project, I provide teachers with the necessary materials, resources, and guidance to establish and maintain hydroponic gardens in their classrooms. To date, I have provided hydroponics kits, children's books, and professional development support to twenty teachers in three preschools and three elementary schools. Robbins Elementary in Trenton, New Jersey, has a large population of immigrant Spanish-speaking students from Central American countries. The teachers I worked with included Natasha Agrawal, an English language teacher.

Hydroponics with Multi-Language Learners

by Natasha Agrawal

Plants growing in our classrooms in the winter? Without soil? Without light? Is it magic? Well, no. It's hydroponics! Professor Ammentorp of The College of New Jersey gave us a taste (quite literally) of growing food without soil and light. Seven classrooms in grades K, one, and two were given books, seeds, nutrients, sponges, and two hydroponic garden–system kits. The tabletop kit included the tank for water and LED lights. Our students planted tiny seeds and witnessed the magic of life as the seed sprouted with green leaves growing bigger in size and number. Instead of soil, there were liquid nutrients to be added to nourish the plants. Within just three to four weeks of planting a tiny seed, our students were able to taste different kinds of lettuce, and cilantro!

Photo 5.2. Tabletop Hydroponics System: Ready to Harvest (Natasha Agrawal)

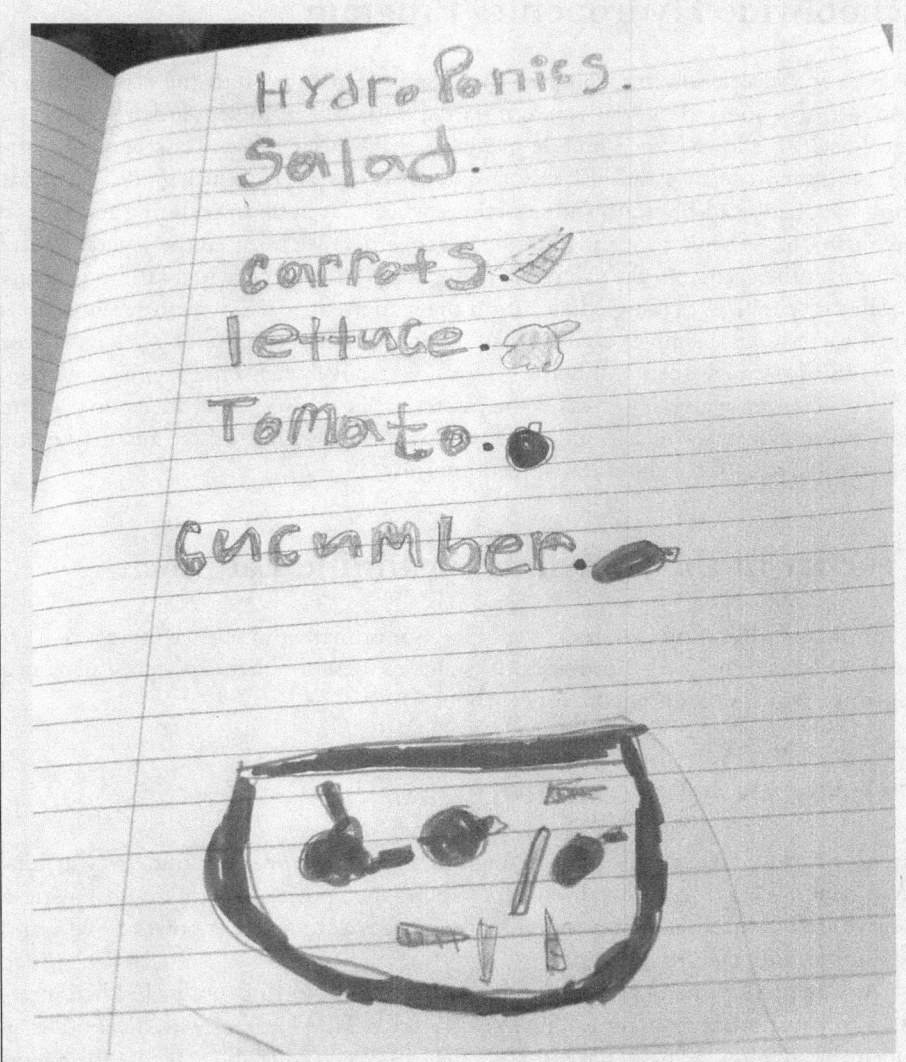

Photo 5.3. Journal Writing with Hydroponics (Natasha Agrawal)

Our students documented the growth, measuring the size daily, drawing diagrams of the shape and size of the leaves. They were scientists—observing, measuring, and documenting. Then they were poets—feeling, thinking, smelling. Before harvesting, they wrote poems about the sensory experience of the fragrance and crunch of the leaves! Finally, they were chefs—tasting, adding, decorating!

It was such a powerful experience for children to witness the growth of a plant from a little seed in a wet sponge. In a world going through climate change and soil erosion, hydroponics is a way to create a sustainable way of producing fresh food. And who knows—the next generation may use this simple technology to reduce our carbon footprint and grow all kinds of food in all seasons, indoors.

Schoolwide Hydroponics Program

At HES we successfully incorporated hydroponic gardening into the classroom and as a part of schoolwide initiatives. We started with several tower garden systems on wheels, which allowed for a mobile plant lab that wheeled from classroom to classroom to support science and SEL lessons and activities, pre-K to fifth grade. We made smoothies, had salad-brations, and sometimes wheeled it in for the calming aesthetic in the classroom. When we expanded to the larger garden, we featured the vegetables and herbs from the garden in the cafeteria. This expansion truly created a farming culture within the school. When ingredients from the gardens are on the menu (even when it is just something as simple as cilantro added into a salsa), the students recognize the freshness of the vegetables and herbs. The students love when mint from the garden, infused in lemon water, is served in the cafeteria. "Featuring" ingredients on a menu that students themselves had an active role in planting, nurturing, and harvesting helped build school collaboration and community.

Teacher in the Limelight: Jeannine Lanphear

In this section, Jeannine Lanphear, the supervisor of math and science (grades 5–8) at Linwood School in North Brunswick, New Jersey, describes her experience starting a flourishing aquaponics program at her school.

STARTING AN AQUAPONICS PROGRAM

We have had an amazing journey getting our aquaponics program underway at Linwood School. We talk with our students about the "power of YET," and having a positive growth mindset. The aquaponics program is a study of the power of YET and methodical, step-by-step adoption.

As supervisor of math and science in grades 5–8 in North Brunswick, I had interest in an aquaponics program for a while. We are a Title I school with a very diverse student population. I knew that our students lack knowledge about where their food comes from, the impact we have on our own food sources, and the science behind photosynthesis and the nitrogen cycle.

I sought to dial students and teachers into the rich, low floor/high ceiling that nature-based learning provides. The push to include climate education in science and social justice in community learning reinvigorated my interest in bringing this approach to the students at Linwood School. However, the right opportunity to leverage it had not come up YET. When the fifth-grade students and teachers came to the former Linwood Middle School as part of an academic restructuring, I realized that we could build eco-activism and nature education directly into the new program for departmentalized science in fifth grade. We began developing the structure for nature and climate education, while I continued to "fish" for a way to make aquaponics happen.

Photo 5.4. Aquaponics System (Jeannine Lanphear)

First, I was able to fund an aquaponics and hydroculture afterschool program upon our return from COVID closure. There were so many activities that our fifth and sixth graders could not do due to restrictions, but this club became quite popular. The teachers did a wonderful job introducing students to fish tanks and care.

When our program received a $500 NJ STEM Month grant for a Title I Parent Academy, I decided to use the funds from NJ STEM Pathways Network to take a baby step instead of waiting for a leap to happen. I purchased materials for the students to develop small take-home hydroponics/aquaponics kits using mason jars as the base for the system. I was also able to purchase a small retail aquaponics tank. The club got busy working with these supplies.

At the parent event, I provided a family workshop, and families decided if they wanted to take home a kit including a fish (for aquaponics) or just the herb (for hydroponics). The kits contained everything that was needed to get past the initial cycling stage. Although most families were nervous to take the fish, there was a huge interest in the idea and a dearth of knowledge about the relationship between the plants, fish, and bacteria involved in the nitrogen cycle. This event and interest became the proof of concept that paved the way for the next step.

At the end of that school year, I had the opportunity to propose how to use some of the pandemic relief funds. I was able to leverage our baby step into a leap of faith. I partnered with Drop the Beet Farms (https://www.dropthebeetfarms.com/) to propose a vertical system. The pictures of the families with their kits, the background of my partner, and recent recognition by the NJ Department of Agriculture

Photo 5.5. Big Fish! (Jeannine Lanphear)

helped me when I gathered the stakeholders to pitch the idea. It was key that the principal, the facilities staff, the grants administrator, and central office administration were involved from the proposal pitch and got onboard to provide the students with this unique opportunity to engage in exciting hands-on science worth coming back to school for.

We located a spot, and some renovations were needed, which were done smoothly in understanding of the need and end goal. We went further and deemed the location the Aquaponics Lab. Getting a window installed in the door was a big win for making the space visible in the school while keeping it distinct. We were so excited when the tilapia came and we saw our first seedlings sprout! Lettuce sales began, as did donations to our local food bank.

Sharing out this success back to NJ STEM Pathways Network resulted in our site being chosen as a feature to kick off NJ STEM Month. A video team came to talk with us and see the lab. It was breathtaking to see our students on the video, projected on a video wall, for STEM Month, and to share how our youngest students feel like scientists in the lab and how students are encouraged to follow interests in nature education. (See the video here: https://drive.google.com/file/d/1nOaZ_ujcxik6Yo2Vlsiht_H_yuWEP-UE/view.)

As luck would have it, the commissioner of education visited our preschool soon afterwards, and she was invited to see our lab. At the Parent Academy, just one year after purchasing our little mason jar kits, we were able to tour parents through the lab. We were on an amazing roll as aquaponics was becoming part and parcel of our school culture and identity while addressing common misunderstandings about photosynthesis, such as models that always show the sun.

Our students understood that photosynthesis can occur in the absence of the sun, and that these models weren't abstracted sufficiently to focus on the real driver of photosynthesis, light energy. They understood that light also means the potential for algae production—for instance, our Linwood Lion's face (an object depicting our mascot that we placed in the tank) began to turn green. Students in the club make observations and take data to understand the phenomena of indoor food production. And they learned to ask really good questions. The club gave us a core group of students with which to try instructional ideas.

Subsequently, the NJ Climate Education grant was released, and I was ready with the idea for our next step—hydroponics. I wanted us to add a large hydroponics system so that our students would be able to compare and contrast the systems and phenomena. We also needed to expand our crop! We were able to add our large hydroponics installation and place it in another public, high-visibility area. We committed to involving as many students as possible and have added a formal aquaponics and hydroculture unit to our fifth-grade science curriculum.

Students are learning about the issues of pesticide use and fertilizer runoff impacting groundwater, and the alternative of natural fertilizer for organic food. Students are caring for living things and developing empathy. Students are giving some of the fruits of their labors to others. Students across the classes rotate through the unit throughout the school year. We are fortunate that we don't have to wait for good weather to cycle a class of students through the related unit!

Photo 5.6. Students Engaging with Aquaponics (Jeannine Lanphear)

Here are the lessons learned:

1. **Take it one step at a time** so that you have proof of interest and concept for the next phase. Science and its application can be intimidating and not clear to all stakeholders. Build everyone's capacity to support your innovation in a step-wise fashion.
2. **Find your people.** When the fifth-grade teachers and students were transferred to Linwood, I discovered that I had inherited teachers with some tank background, a lot of environmental interest, and a real passion to get the students excited about nature education. Get to know your staff and what they love. Teaching what you have passion for makes learning infectious. I could not have been successful asking teachers who have a low tolerance for "ick" to work with fish, algae, and tanks. NJ STEM Pathways Network was there before and after the grant as STEM cheerleaders to celebrate every STEM step. The relationship with our vendor has directly benefited our students, as they have engaged with our students, showing our students that, yes, you can do this for a career. Don't forget that your people are not just your science colleagues. Everyone in proximity is also your stakeholder. The facilities staff, principal, and central office administrators have been instrumental in the success and promotion of this approach to nature, science, climate, sustainability, social justice, ecology, and engineering education.
3. **Seek synchronicity.** Gather your ideas and look for the wave that will help carry the idea to fruition. I was at Linwood for seven years before I pitched this idea. The right staff, funding, and the imperative for climate education all made the difference. There are many great ideas out there. Find the match to the landscape, the right idea for the right time, to bake-in success.
4. **Keep it real.** Make sure your project isn't just tinkering with STEM and leaving unfinished learning on the table. This project's goals were based on the three-dimensional standards for learning in science. Staying true to the teaching and learning at hand was an easy way to keep the end in mind for the vision of the project and for decision-making along the way. We continue to strengthen curricular connections while providing many of our students with experiences and exposure that would not have been in their reach.
5. **Attend to the small things, as they send a BIG message.** When we messaged through signage and adult talk that the space is no longer a prep room and is now an Aquaponics Lab, the kids believed us. They knew it was a special place where science learning happens, not just a fish tank. The students will readily tell anyone that fact. Student access, in the fifth grade, to a place called a lab, where they care for living things and make observations and collect data, is positively impacting student STEM identity. When we put the observation window in the door so that anyone in the hall sees the installation, lettuce and tilapia became a part of the school's identity. Adding a hydroponics installation at the window to the street signals that our school appreciates nature's bounty.

I have my next step cued up, and I am waiting for the right time to "spring" it. I think we are getting close!

Recommended Websites

- Drop the Beet Farms https://www.dropthebeetfarms.com/
- Green Our Planet https://www.greenourplanet.org/hydroconnect
- Hydroponics Online http://www.hydroponicsonline.com
- Kids Gardening https://kidsgardening.org/?s=hydroponics
- San Diego Hydro https://sdhydroponics.com/2012/05/11/how-to-grow-hydroponically-overview-of-grow-systems/
- Lifelab www.lifelab.org
- The Edible Schoolyard Project www.edibleschoolyard.org
- Tower Garden https://www.towergarden.ca
- Greenplur Mason Jar Aquaponics https://greenplur.wordpress.com/aquaponics-fish-garden/
- Go Green Aquaponics https://gogreenaquaponics.com/blogs/news/guide-to-mason-jar-aquaponics
- Green Bronx Machine (Stephen Ritz) https://greenbronxmachine.org/
- Ponics Area Mason Jar Hydroponics https://ponicsarea.com/mason-jar-hydroponics/

CHAPTER 6

A Healthy Harvest

CULINARY ARTS IN THE CLASSROOM

Among the unfortunate consequences of the emphasis on "academic achievement," "rigor," and high-stakes testing in the 1990s and 2000s was the elimination of home economics and other life skills and vocational education programs in schools. Due to funding cuts, school kitchens and woodshops were removed from schools, along with their corresponding curriculum. This was a huge mistake. There are so many benefits of incorporating food preparation and cooking into student learning. Using culinary arts and sciences in the classroom provides learning opportunities for students and teachers alike.

Students love to eat what they grow and cook. Cooking experiences can be integrated to support social-emotional development and health and wellness. For instance, cooking supports student self-awareness by exposing children to various foods, helping them discern their taste preferences, and reducing negative associations with disliked foods (Brouwer & Neelon, 2013). Additionally, cooking promotes mindfulness and appreciation of the food that students have created (Russell, 2011). It is important to note that low-income students, who often lack access to fresh food in their communities, especially benefit from opportunities for healthy snacks and meals (Cunningham-Sabo & Lohse, 2013; Ozer, 2007). Further, cooking is hands-on and interdisciplinary. It can be easily and authentically incorporated into each of the major content areas.

Bringing cooking into the classroom doesn't have to involve an oven or expensive equipment. We include ideas in this chapter that can be done on a small budget, such as taste testing, as well as those for more established programs such as at HES where, through persistence and patience, Helen was able to get a demo kitchen installed. We encourage teachers to take the positive feedback (there will be a lot!) and build on that positivity to try even more.

Incorporating herbs and vegetables grown in school gardens provides yet another level of student participation. Hydroponics and container gardens are an easy way to grow vegetables and herbs if a school doesn't have an outdoor garden (see chapters 4 and 5). This chapter provides ideas for planned use of the harvest, as well as ways to use specific herbs or vegetables when unexpected volume occurs.

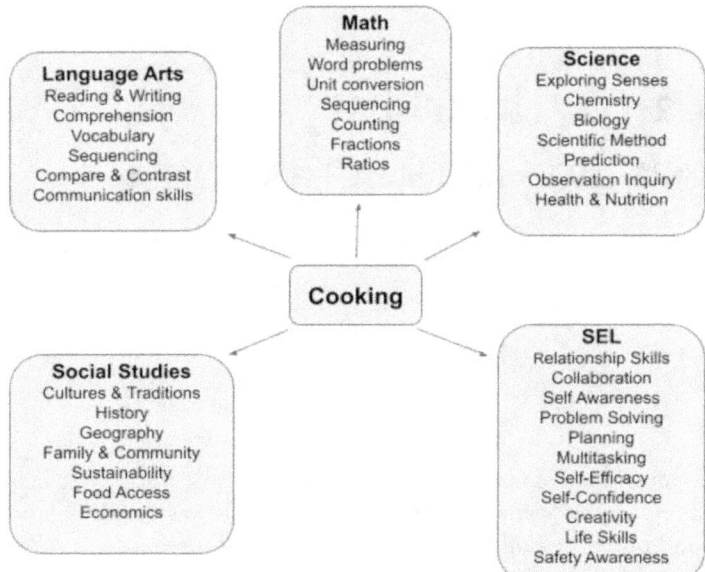

Figure 6.1.   Integrating Cooking across the Content Areas

Connecting to Culture and History

Students can learn about cultures from all over the world through food, especially vegetables and herbs. They can research a culture and learn about the herbs and vegetables grown and eaten there. Through researching certain plants they can learn about geographical areas with similar and different climates to their own. They could look at a plant such as garlic or food such as bread and their use in cuisines all over the world. And they could look at the history of the food item's trade. They can also study a country's cuisine, such as Italian pesto with basil; Greek tzatziki with cucumber, yogurt, mint, and/or dill; or Indian curry.

Parsley is a great example of an herb that is very easy to grow, and in a hydroponics system it is sometimes in overabundance. This can be an opportunity to learn and experience different cultures. For example, tabbouleh is a Lebanese salad made primarily of parsley. It also has mint, lemon juice, olive oil, tomatoes, and bulgar wheat (optional). Chimichurri is an Argentine sauce made with parsley that is traditionally used as a condiment for meats or fish but can be used as a spread on bread or a salad dressing. It includes parsley, garlic, olive oil, vinegar, and oregano. Trying gluten-free breads or vegan breads with different spreads also deepens knowledge and understanding of foods in a diverse classroom.

There are many literacy connections that can be made with food. Books such as *Everybody Cooks Rice* and *Bread, Bread, Bread* show the different ways people around the world prepare and eat similar ingredients. There are also many examples of folktales that involve herbs and vegetables, from Jack and the Beanstalk and the Little

Red Hen to the role of garlic for protection against vampires and other supernatural entities. Tomie dePaola has many wonderful books for children that involve food, such as *Strega Nona*. There are numerous Native American stories about planting and food, such as the story of the three sisters (corn, beans, squash) and *Fry Bread* by Kevin Noble Maillard, a beautiful book that also depicts the history and impact of colonization in a developmentally appropriate way for young children.

> **Children's Books**
>
> *Maddie's Fridge* by Lois Brandt
> *Pumpkin Soup* by Helen Cooper
> *Thunder Cake* by Patricia Polacco
> *Bring Me Some Apples and I'll Make You Pie* by Robbin Gourley
> *Loaves of Fun* by Elizabeth Harbison
> *What's on Your Plate?* by Whitney Stewart
> *I Will Never Not Ever Eat a Tomato* by Lauren Child
> *Strega Nona* by Tomie dePaola
> *Yasmin the Chef* by Saadia Faruqi
> *Little Pea* by Amy Rosenthal
> *Bee-bim Bop!* by Linda Sue Park; illustrated by Ho Baek Lee
> *The Seven Silly Eaters* by Marla Frazee
> *How to Feed Your Parents* by Ryan Miller
> *Cora Cooks Pancit* by Dorina Gilmore
> *My Food, Your Food, Our Food* by Emma Carlson Berne
> *Fry Bread: A Native American Family Story* by Kevin Noble Maillard
> *Same, Same but Different* by Jenny Sue Kostecki-Shaw
> *Everybody Cooks Rice* by Norah Dooley
> *Compost Stew: An A to Z Recipe for the Earth* by Mary McKenna Siddals
> *Bread, Bread, Bread (Foods of the World)* by Ann Morris and Ken Heyman
> *Comiendo el Arcoíris—Eating the Rainbow* by Patricia Barrera Boyer

Food offers an opportunity to create family and community connections by inviting guest chefs or parents from different cultural backgrounds to conduct interactive cooking demonstrations. This allows children to observe and participate in the preparation of dishes using herbs and vegetables. Include discussions about the cultural heritage of the recipes and the importance of specific ingredients. Children can develop a greater appreciation for diversity while also gaining practical knowledge about the role of these ingredients in global cuisines.

Exploring Greens

TASTE TESTING

Introducing young learners to the idea of eating "green" can sometimes require a gradual approach. Some children may be resistant to trying new things. However, when they witness their classmates eating and trying vegetables (especially if they have planted, observed, and nurtured the plants) even the most hesitant eaters often transform into enthusiastic advocates for eating healthy. Conducting taste tests is an exciting activity for students of all ages, and there are multiple ways to carry them out.

A great place to start is simple taste tests, for example, trying a few types of lettuce with a simple dressing like oil and vinegar, or lemon juice and olive oil, or with herbs such as cilantro and dill. For schools with hydroponic systems, kale is a lettuce that thrives! We provide students with the opportunity to taste it in three different preparations: raw, roasted with olive oil and kosher salt, or sautéed with fresh lemon and pepper. This approach allows students to explore a new vegetable in various forms, enabling them to recognize that while they may not like raw kale, they might enjoy it when cooked. It encourages them to consider their senses of smell and taste while also practicing positivity when expressing their opinions about food and their peers' choices.

Herbs are an exciting way to explore smell and taste. Lemon balm, cilantro, and mint have strong smells and flavors that students enjoy. Comparisons between the texture and taste of dried and fresh herbs is an engaging activity for students. Spice containers filled with dried or fresh herbs can hone observation skills as students describe the smell and memory skills as they guess the spice.

SMOOTHIES

Green-smoothie brain breaks are a wonderful way to create a healthy routine, especially if you have fresh produce from hydroponics or a school garden. This practice serves as a gentle introduction to nutrient-rich greens like spinach, kale, and romaine lettuce. To enhance the taste while preserving the vibrant green color, combine the greens with apple, pear, and/or pineapple juice along with coconut water. For those feeling adventurous, add avocado to create a velvety texture in the smoothie. As part of the experience, you can provide students with a food "altar" where they can choose from a variety of wholesome ingredients for their personalized small-batch smoothies. This empowers them to take control of their dietary choices while utilizing the vegetables they have cultivated. It also helps them develop a habit of making healthy food choices. However, as with all food in schools, it is important to note that any specific dietary restrictions or allergy concerns must be carefully considered and checked before consuming food in the classroom or cafeteria.

IT'S A SALAD-BRATION

At HES, students have the opportunity to handpick lettuce and herbs from our hydroponic garden to create their own custom salad dressings. We encourage children to bring their favorite salad toppings, adding a personal touch to the harvest. This joyful celebration of their hard work and dedication to nurturing the garden is known as a Salad-bration! The Salad-brations have become highly anticipated

Photo 6.1. It's a Salad-bration! (Helen Corveleyn)

events, so much so that a first-grade class recently opted for a Salad-bration as their class reward instead of a traditional pizza and ice cream party. The sense of community that develops around the food table during these gatherings is truly remarkable and shouldn't be underestimated.

Food Preservation

Food preservation extends the shelf life of food and prevents spoilage. There are many types of traditional and modern preservation methods, including: canning, pickling, drying, freezing, fermentation, smoking, salting, sugaring, vacuum packaging, and using root cellars. Learning about food preservation is a useful life skill for students, and lessons can connect to numerous Next Gen Science Standards, particularly those related to Energy (PS3). There are many social studies connections related to food preservation as well. Students can research traditional methods of preservation historically and in different cultures and climates. It is a great opportunity to visit farms and/or living museums or for a visit to someone to teach students about food and preservation. That person could be a family member or a person from the community, or it could be an online presentation. Some of the preservation techniques are easier and safer than others to bring to the classroom. In this section, we discuss just a few. There are many great resources for lesson activities, such as the *Farm to School* and *Agriculture in the Classroom* websites as well as books.

Websites

Edible Schoolyard Project https://edibleschoolyard.org/
Farm to School Network http://www.farmtoschool.org/
Agriculture in the Classroom https://agclassroom.org/
TeacherVision: Food https://www.teachervision.com/resources/food
Cooking with Kids https://cookingwithkids.org/
Food Literacy Center https://www.foodliteracycenter.org/

USDA

- Team Nutrition Cooks USDA https://www.fns.usda.gov/tn/cooks
- Snap-Ed—Nutrition Education USDA https://snaped.fns.usda.gov/snap-ed-works/nutrition-education

Kids Gardening

- Fermenting https://kidsgardening.org/digging-deeper-fermentation/
- Abenaki Food and Heritage https://kidsgardening.org/program-spotlights-abenaki-food-and-heritage/

Pickling

Introducing children to pickling with vinegar is a simple and enjoyable way to incorporate cooking into the classroom. Pickling cucumbers is an excellent way to begin. Cucumbers can be grown in a school garden or purchased, and dill thrives particularly well in a hydroponics garden (see chapter 5). Pickling offers many STEM connections such as:

- **Observing, predicting, and exploring flavors:** Students can engage in predicting and investigating how different spices impact the flavors of pickles. For instance, they can explore the effects of adding more garlic or experiment with pickling various vegetables. At HES, students were excited to discuss the possibilities and potential variations in flavors.
- **Chemistry and pH:** Pickling provides a hands-on experience for students to observe chemical changes and learn about pH levels. By comparing fresh cucumbers to pickled cucumbers, students can explore how an acidic solution such as vinegar transforms the cucumber. This activity aligns with NGSS Core Idea PS-1, Matter and Its Interactions, for second and fifth grades.
- **Mathematical connections:** Mathematics can be integrated by graphing preferences, such as who prefers cucumbers versus pickles, spears versus coins, and so on. Other mathematical concepts that can be explored include making comparisons, measuring, estimating, and even adding and multiplying fractions to adjust the quantities for pickling cucumbers.

Photo 6.2. Pickling Preparation (Helen Corveleyn)

In our article "The Promise of Pickles: Hydroponics in the Classroom" (2020), we describe the exciting ways Helen used pickling at HES. For example, third-grade students explored pickling as a sustainable product, discussing the reuse of glass jars and the cost-effectiveness and environmental benefits of growing vegetables hydroponically or in soil gardens. The ideas flowed freely. Student comments included:

- "What if each class chose a different vegetable to grow and we pickled them to sell at a community farm stand?"
- "Would different vegetables have different prices? Could we charge more for organic vegetables?"
- "We could design our own labels!"
- "If we called them 'gourmet,' could we price them differently?"
- "More garlic means spicier pickles. We could offer 'Mild,' 'Medium,' and 'Wowzers! That's Spicy' options."
- "This could involve the entire school. It's so much fun that nobody would want to miss out."

These examples show how growing and eating their own food engages and motivates students. Further, students become immediately invested when *their ideas* take off and they get to investigate and take a deeper dive into a community-driven idea. Let the farm stands begin!

Dehydration

In addition to pickling, dehydration is another preservation technique that can be brought into the classroom. Dehydrators can be bought for as cheap as forty dollars. A store-bought dehydrator comes with several trays so groups of students can each create a tray of items to dehydrate. Just like pickling, dehydration is a wonderful opportunity to have taste tests and discuss taste preferences. Social studies connections can be made to ways Native Americans and other people historically preserved food. Students can make predictions and observations about dehydrated food and learn about water and moisture in fruits, vegetables, herbs, and flowers. Comparisons of size and weight before and after dehydration make for authentic math connections! Using solar heat, or a hanging air dehydrator, are other options for dehydration, especially for herbs and flowers. It is a great experience to compare the methods of dehydration.

Vegetables and fruit should be cut up prior to putting them in the dehydrator. This is also an enjoyable activity for the students. Dehydration, even in a dehydrator, does take many hours, so it is best to start it first thing in the morning, or plan it as a two-day lesson. Fruit and vegetables to dehydrate include: apples, bananas, strawberries, pineapple, mangos, kiwi, watermelon, pickles, sweet potatoes, tomatoes. Also fruit leather is a fun snack to make. There are recipes online, but basically it involves pureeing fruit, adding some applesauce, and placing it on a fruit leather tray in the dehydrator. Fruit such as apples and bananas need lemon juice so they don't brown, which is an interesting phenomenon to explore!

Herbs and edible flowers can easily be dehydrated. These can be used for teas or in other recipes. If possible, students can forage for edible plants such as dandelions, chickweed, and wild violets. However, often school grounds are sprayed with pesticides, so this may not be possible. Some herbs great for dehydrating include: basil, dill, mint, cilantro, rosemary, thyme, parsley, oregano, and greens (such as kale).

Teas

Most herbs can be made into tea, with some having a stronger taste. A tea garden can be purposefully created outside in a bed or container, or in the classroom. Teas are also a great way for children to compare the taste of fresh herbs to when they are infused in hot water. It is usually better to use fresh herbs for tea! This sometimes is not realized since people are used to buying tea. It is also interesting to compare dried herb tea to fresh herb tea, especially if students can dehydrate the herbs themselves. Always check if there are recommendations for children. Anything in the mint family makes for a flavorful tea: peppermint, spearmint, lemon balm, basil, lavender, rosemary, as well as other herbs such as parsley, dill, and lemon verbena. Berries, flowers, and many tree leaves make for great tea, for example, dandelion, chamomile, magnolia, and pine.

Sampling hints: It's always a good idea to have students sample the tea in stages. Have students taste the tea without anything added first. Encourage students to taste and really reflect on how it feels in their mouth and how it smells. Praise them for trying something new. From there you can add different flavors: lemon, honey, or milk. Typically students will enjoy adding any of these combinations for the mere thrill of trying something else. This taste testing encourages the joy of stepping outside your comfort zone to try something new.

Book Corner

Growing Vegetable Soup by Lois Ehlert is a fantastic book for young children, showing them the farm-to-table aspect of cooking. At HES, we asked kindergarten families to contribute one veggie that their child wanted to bring in from their fridge. We emphasized that they didn't have to buy anything; whatever they had, they could share, and we would make into soup. We could also add a few things from the garden to balance out what was brought in. For students who didn't bring in a veggie, we let them choose from the garden at school. We had many inquiry-based discussions develop from these surprises! They included:

- What vegetables grow in the dirt and need to be washed? Does the dirt make the water change?
- What vegetables have seeds and which ones only have leaves?
- Do vegetables float or sink?
- Do vegetables have layers? Do vegetables come in parts?

This book and activity develop student vocabulary and increase food literacy. Each child was so proud to contribute their "one thing" (something about a singular item is very approachable for a kindergarten student). After we cut, sauteed, and added salt, pepper, and broth to the soup, they had voracious appetites! They LOVED the soup! We put it in a crockpot after cooking it on the stove to bring to the classroom. This gave us time to let it cook for a bit, but it also allowed the kids to really look at the soup cooking in the crockpot after it was on the stove. We reread the book after making the soup and the kids had a great time saying, "We did that," after each page!

Schoolwide Programs

VITAMIX AND THE POWER OF SOCIAL MEDIA

At HES, we wrote a grant and bought a Vitamix blender that we used for smoothies and recipes with every age level. Whenever we posted on social media, we always tagged Vitamix. We were delighted when Vitamix consistently responded. When they rolled out their food cycler—a tabletop food composter—we tweeted them, and Vitamix donated a food cycler to our program to learn about food waste, help combat climate change, and help understand composting—a new twist on the "tech" part of STEM in the kitchen!

AFTERSCHOOL TOP CHEF PROGRAM

The principal at HES conceived the idea of creating an afterschool program inspired by the TV show *Top Chef*. Chefs in the area came to the school to teach kids a signature dish from their restaurants. Students prepped the food to cook with the chef. Some favorite dish preparations included four-cheese ravioli from scratch, grinding hamburgers from sirloin and brisket, making fresh sushi, and enjoying spicy street tacos! One year, as a conclusion to the program, the students from the club helped the cafeteria to prepare a lunch for the entire school to see what lunch preparation included. Equipped with hairnets, gloves, and aprons, they chopped veggies, prepared a fresh salad bar, and made homemade soup. Students couldn't believe the amount of work that it took to prepare a homemade meal for all the students in the school. They also developed a deeper understanding of food waste and got upset when they saw students not finishing everything on their plate. In reflections on the activity, they expressed their shock at all the "waste," both of food and time.

BUILDING A DEMO KITCHEN

After creating a hydroponic garden in the school and beginning to cook with it in the cafeteria, we decided a kitchen right across from the vertical garden would be an amazing opportunity for growing and cooking food. There is nothing fresher than ingredients that are handpicked exactly when you need them in a recipe! During COVID we (Helen and HES principal David Friedrich) created a cooking show for families to follow along at home. This was a wildly successful show. We had a Zoom link for families to join in—and we even made vegetarian side dishes for Thanksgiving in the month of November! It was a simple setup, using our school computers to record the show. Kids and their families got such a thrill from cooking along with us. They were given the recipe and ingredients ahead of time, so they were ready to cook and follow along. We made cranberry sauce, cornbread, brussels sprouts with meat-free bacon, and applesauce. It was a huge hit. Many children contributed their own recipes and side dishes for their family's Thanksgiving dinner!

Herb Biscuits Recipe

Photo 6.3. Making Dill Biscuits (Helen Corveleyn)

Herbs can be infused into dough for a yummy treat! Making and eating bread has many SEL benefits, it is a wonderful opportunity for conversation and camaraderie. Creating a herb bread from scratch requires several rounds of kneading dough. Kids love kneading dough, and while biscuits can become tough instead of flaky during this process, we find that in the end, it really doesn't matter! At the heart of cooking is the process and shared experience kids bond over. The physical activity of making dough develops motor skills and is a calming activity. When faced with an abundance of any herb, such as dill, incorporating it into biscuits is a simple but satisfying way to use ingredients.

(continued)

INGREDIENTS

2 cups all-purpose flour
3 teaspoons baking powder
2 to 3 teaspoons dill weed
¾ teaspoon salt
¼ teaspoon pepper
½ cup cold butter
2 eggs, lightly beaten
½ cup plus 1 tablespoon half-and-half cream, divided

Directions: In a large bowl, combine the flour, baking powder, dill, salt and pepper. Cut in butter until the mixture resembles coarse crumbs. With a fork, stir in eggs and ½ cup cream until just moistened. Drop ¼ cupfuls 2 inches apart onto an ungreased baking sheet. Brush tops with remaining cream. Bake at 450° for 10 to 12 minutes or until golden brown. Serve warm.

Optional: Make homemade butter! Making butter with students is always a hit. Simply put heavy cream in glass or plastic jars with a sealed lid and shake it up. They love to see the cream transform into butter. However it does take a lot of shaking up—about 5 to 10 minutes. Adding finely chopped herbs makes it even better! Herbs such as parsley, dill, basil, sage, rosemary, thyme, and so on, all make for delicious butters.

CHAPTER 7

Life Is Short, Play in the Dirt
EARTH SCIENCE, PHYSICAL SCIENCE, ENGINEERING, AND MATH

Teaching STEM in and with nature engages students in hands-on, experiential learning while fostering a deeper appreciation for nature. Using the outdoors as a model and inspiration taps into students' sense of wonder and curiosity. It also engages them in the process of creating solutions to problems. In this chapter, we focus on ways to integrate nature into STEM teaching, with a focus on engineering, physical science, earth science, and math.

Exploring Nature's Solutions to Problems

It is exciting to engage students in discussions of how nature solves a lot of problems on its own and that human solutions to problems often come from observing nature. A powerful introduction is to walk around school property with the question, "How does nature solve problems?" There are examples everywhere. For example, weeds coming through sidewalk cracks are solving a problem. Morning glories or other vine-like plants can be found twisting and winding up a fence. Vining plants are a wonderful opportunity to ask students questions such as, "What problem is nature solving by having a stem that is so flexible and thin?" This is a great way to introduce biomimicry, which is taught in first grade as part of the Next Generation Science Standards. It is also an exciting way to teach seed dispersal. Helicopter seeds from a maple or the barbs on the seeds from sweetgum trees are fascinating examples of "tree technology." Seeds need to get far away from their tree, and so nature invented a perfect, persistent method of the seed traveling onto a furry friend out into the world!

There are so many examples of humans using nature to create innovative designs that can be shared with students or that students can find through their own research. Examples include the kingfisher as the model for a high-speed train or how the gecko's sticky feet translate into mountain climbing equipment. A wonderful way to teach biomimicry and engineering is to solve a problem on the playground or in the cafeteria using nature as inspiration. Here is an example aligned with the first-grade NGSS 1LS1-1 (*Use materials to design a solution to a human problem by mimicking how plants and or animals use their external parts to help them survive, grow, and meet their needs*).

Begin with this scenario:

> There is an open garbage can outside, right next to our playground. Since we sometimes eat snacks outside, small bits of plastic wrappers often get trapped in hard-to-reach spaces. How could we build a device to pick up the garbage using a solution inspired by nature?

As they brainstorm, students may think of animals such as an elephant: a long trunk with a "finger" at the end, that is flexible but firm, long enough to reach into hard-to-reach places and grip delicate pieces of small materials. They may think of a gecko: using sticky removable materials to adhere to any surface (with the exception of Teflon!). Geckos can also stay vertical for a long period of time and scale vertical surfaces. A gecko-inspired solution would allow any type of garbage to be picked up from the playground whether it was on the ground or stuck to a structure. They may consider a Venus flytrap (always a favorite!). As the fronds of a Venus flytrap close when triggered by tiny hairlike structures located on the inside of the hinged part of the plant, students may design a similar structure using a trigger to close over the trash to lock it in. Students can design and build a prototype of a device modeled after plants or animals (the best for deep understanding and transfer of knowledge), or they can draw and label their designs (this needs fewer materials and can still be very effective, but does not connect as deeply as actually building the model).

Podcasts: Learning and Communicating about Science!

Through podcasts and other digital media sources students can be exposed to stories from all over the world. Podcasts support listening skills and focused attention on specific topics regardless of reading level. With a simple search there is a wide variety of podcasts for different ages, skill levels, and topics. This makes a great "station" activity. Podcasts can support differentiated instruction by pairing a short podcast with an article-reading station, a group workstation, and a building station. Another successful use of podcasts is during snack time. Teachers can turn on a podcast to listen to as students eat their snack, and students typically find it quite relaxing.

Many classrooms have also incorporated podcasts as an assessment product after a unit of study. Creating a podcast includes researching the topic, writing a script, working with a partner to play the expert and the guest, and then recording the podcast (some schools have podcast recording studios). An easy recording device is WeVideo. This current and relevant source of learning information is a motivator to older students who may enjoy following podcasts in their free time and can be a great way of engaging learners with special needs.

Resources for Using Podcasts in the Classroom

Teach Create Motivate https://www.teachcreatemotivate.com/podcasts-in-the-classroom/
Reading Rockets https://www.readingrockets.org/article/creating-podcasts-your-students
Podcast design sites: Camtasia and Anchor

> **Great Podcasts for Students**
>
> - *Wow in the World*: NPR podcast for kids; high-interest science and engineering topics for K–5
> - *The Show about Science* highlights science careers, hosted by a 7-year-old
> - *Earth Rangers*: A variety of topics appropriate for K–4
> - *Tumble*: Extensive science content
> - *But WHY?* General topics that serve curious kids
> - *The Big Fib*: A game show where children have to figure out the truth from the lies
> - *Tai Asks Why*: An 11-year-old asking big questions and seeking answers
> - *The Ten News*: News for curious kids

Engineering

With the adoption of NGSS by many states, the emphasis on engineering became new territory for many teachers. Engineering, however, offers exciting opportunities to connect with nature. Students can learn how engineers draw inspiration from nature to solve problems and create innovative designs. Engineering challenges can be directly inspired by natural phenomena, such as in the example above of designing devices based on animal adaptations or finding solutions to environmental issues. By providing students with natural materials like sticks, stones, and leaves, their creativity can flourish as they construct simple structures and test them outdoors.

> **Building Boats and Bridges**
>
> Children love to build and, given the time and opportunity, will be enthusiastically engaged as they create and test innovative designs. Bridges and boats are especially exciting and full of learning potential. Students can use sticks, mud, string, and leaves as building materials. This takes the sink-and-float activity to a new level, and makes it more authentic. Especially for younger children, integrating stories and toy animals and humans makes the activity even more engaging.
>
> Jessica Jozak, a preservice-teacher student in Louise's program, created a STEM challenge with students in Helen's class. Students designed, created, tested, and redesigned rafts made from natural materials, first using a bin in the class and then a nearby stream. Students loved it!
>
> Adding constraints makes the activity adaptable to different grade or developmental levels. Constraints can include:
>
> - Material limitations: Challenge students to build their structures using a specific number or type of natural materials.
> - Size: Set size limitations for the structures. Students may need to design a bridge with a maximum length or height.
> - Stability: Specify that the structures must be able to withstand a certain amount of force (extreme weather such as rain or wind) without collapsing.
> - Aesthetics: Provide guidelines for the overall look and appearance of the structures.
> - Function: Specify a purpose for the structures. For instance, students may need to design a boat that can float and carry a small load.
>
> *(continued)*

Photo 7.1. Building a Raft (Helen Corveleyn)

Photo 7.2. Testing the Raft in the Stream (Helen Corveleyn)

Eco-Homes: Children love to build and create with traditional building materials such as Legos, K'Nex, Keva blocks, and Magna-Tiles. As they get older (fourth through sixth grade), they may not have opportunities to build with these materials in school, however there is so much value and creativity that comes with it when paired with the right prompt. When you begin to look at eco-design and eco-architecture, drafting and building an eco-conscious home is easily represented through manipulatives. Of course you can have them create a blueprint or an architectural plan, but the fun and engagement comes when they are actually doing the building. Successful prompts may include:

- Create an eco-home that has three energy-saving features built in
- Build a home where the main feature is reducing carbon dioxide in the environment
- Build a home that significantly reduces daily water usage
- Choose a state in the United States that is conducive to a certain type of alternative energy source like wind, solar, or geothermal
- Look for an architect in your area to come in and talk about green infrastructure

Gnome Homes/Elf Houses/Fairy Dwellings: Children love designing and constructing small dwellings or habitats for imaginary creatures like gnomes, elves, or fairies. Students can use natural materials such as twigs, leaves, bark, and pebbles to create whimsical and intricate structures.

Rock Sculptures: Have students collect rocks and explore balance and stability by creating rock sculptures. They can compare the experience of building on different surfaces, such as flat and hard (concrete/sidewalk), soft, and/or varied, such as dirt, sand, and grass. They can learn about and view the work of environmental artists such as Andy Goldsworthy.

Feeder Challenge: Students can be challenged to create devices that hold seeds or food for animals. They can experiment with different shapes, sizes, and materials to create a structure that can withstand weight and weather conditions. Students can document their designs, take pictures, and share their creations with the class.

Engineering Snowflakes: By utilizing technology tools like Tinkercad, students can explore the intricate patterns and symmetry found in snowflakes. They can create and manipulate digital models of snowflakes, experimenting with various combinations of branches and shapes to mimic the beautiful symmetrical designs found in nature.

Leave It to the Beavers!

An engaging, hands-on engineering activity is building a model beaver dam. Beavers are fascinating creatures and there are many books and videos to support student learning. Younger children can build dams and learn how beavers impact their environment to meet their needs. This is an engaging activity that aligns with NGSS for Kindergarten (K-LS1-1 and K-ESS2-2). Dams can be built outside or in aluminum or plastic bins in the classroom. Natural materials such as mud, sticks, and leaves make the activity authentic. Adding pretend beaver families and other animal representatives and creating stories will make it exciting for the younger students. It is a great collaborative activity as students work in groups to create a dam and tunnels. Older students can map and design larger dam and tunnel systems. They can also research and learn about beavers' role in the ecosystem as an ecological indicator and as a "climate change hero."

> **Beaver Resources**
>
> - *Build, Beaver, Build! Life at the Longest Beaver Dam* by Sandra Markle
> - *Animal Engineers Beaver Dams* by Nancy Furstinger
> - Beavers as Keystone Species https://www.beaversolutions.com/beaver-facts-education/beaver-benefits-a-keystone-species/
> - Beavers as Ecosystem Engineers and Fight Climate Change Too https://environmentamerica.org/articles/beavers-are-ecosystem-engineers-and-fight-climate-change-too/
> - Beaver Emerges as Climate Solving Hero https://www.scientificamerican.com/article/the-beaver-emerges-as-a-climate-solving-hero
> - Video about beaver research in Utah https://www.youtube.com/watch?v=2_cml_cXPmE
> - Beavers and Wildfire: a stop-motion story by Emily Fairfax https://www.youtube.com/watch?v=IAM94B73bzE
> - Emily Fairfax researcher video https://emilyfairfaxscience.com/
> - National Geographic beaver video https://www.youtube.com/watch?v=jBaXlyq3E8M/

Physical Science

Integrating the study of physical science with real-world experiences in nature provides students with hands-on opportunities to observe, investigate, and understand the fundamental principles that govern the physical world. The NGSS outlines four core ideas in physical science: PS1—Matter and Its Interactions; PS2—Motion and Stability: Forces and Interactions; PS3—Energy; and PS4—Waves and Their Applications in Technologies for Information Transfer. In this section, we share a few ideas aligned with each of the core areas.

MATTER AND ITS INTERACTIONS (PS1)

There are so many ways to engage students in hands-on activities that explore the properties of matter using natural materials. They can investigate and compare the ways different objects act on water versus land by experimenting with floating leaves, twigs, or rocks in water. Students can also compare the physical properties of different types of soil, such as texture, color, and permeability. An outdoor inquiry lesson can explore what materials are used (and why) for things like playground equipment, garden fencing, birdhouses, animal feeders, and outdoor shelters. Teachers can create a table with a few different structures listed and have students notice characteristics of the outdoor materials—this is far more authentic than having students identify characteristics from pictures in a book.

FORCES AND MOTION (PS2)

The outdoors provides many opportunities to observe and investigate forces and motion. Students can learn about concepts like gravity, friction, and the effects of forces on objects. They can explore how wind affects the movement of objects such as leaves and flags. Students can also experiment with rolling objects down inclined surfaces (like slides or playground structures versus a hill or asphalt ramp) or observe how rocks or seeds are carried away by flowing water.

Nature mobiles: A collaboration with the art teacher

Helen developed this lesson in collaboration with the art teacher. The lesson focuses on the concept that when balanced forces act on an object, it remains motionless, but unbalanced forces cause the object to move toward the greater force. Science unites with art through the creation of mobiles—a perfect opportunity to explore visual and physical balance together. To begin the lesson, the art teacher showcased examples of famous paintings that exemplified symmetry and balance on a canvas. For the STEM component, we engaged the students in kinesthetic activities to demonstrate how balance is achieved. They practiced standing on one foot, two feet, and holding objects of equal weight in each hand. Additionally, they explored the concept of balance by using popsicle sticks or pencils as a fulcrum and experimenting with different spots for finger placement. The simplicity of these activities captivated the students and allowed them to experience balance firsthand.

After examining balance through art and physics, we introduced the project of creating a nature mobile. This activity brought nature indoors and provided students with the space to unleash their creativity, whether they opted for a simple or intricate design. If time permitted, we encouraged the class to go outside and collect a variety of small sticks, interesting grasses, and even flowers (reminding them that flowers would eventually dry out and change in appearance). However, if time was limited, we provided a collection of materials for them to use.

Activity directions:

1. *Choose a color palette for the mobile.* Artistic balance can be incorporated by suggesting that students paint one half of the stick with a primary color and the other half with a complementary color. Students can select the yarn or twine to be used for attaching the mobile.
2. *Create the top of the mobile.* This can be a simple shape like a triangle or rectangle, or the teacher can introduce three-dimensional geometric shapes (which also ties in with math concepts). For older children, the three-dimensional shape provides an opportunity to demonstrate balance by attaching the string to different parts of the shape. For younger children, a flat shape may be more suitable.
3. *Begin experimenting.* Add different sticks to attach to the mobile. Encourage students to explore sticks of various lengths and thicknesses to add visual interest.
4. *Tie it together.* As students start tying their creations together, encourage them to test the balance of the mobile by continuously holding or hanging it. A few clotheslines in the classroom is a great way to help them to test and adjust their creations.
5. *Final embellishments.* Invite the students to decorate the mobile, ensuring that they consider balance even with the addition of new elements.

ENERGY (PS3)

In chapter 6 we discuss lessons related to food preservation that align to the concept of energy. Solar ovens are an exciting way to engage students, even if it's just heating chocolate and marshmallows to make s'mores! There are many great ideas on STEM websites about how to build ovens, but the main idea to keep in mind is that solar ovens use the sun's energy to warm the air molecules around the food. Thus, the heat must be trapped and retained by some sort of insulation for the slow cooking of the object in the oven. Because the cooking happens slowly, moisture is retained and heat is captured for the food inside. A great base to use is a pizza box, as it is already "hinged" and ready for building. Providing insulating materials such as Bubble Wrap, upcycled Styrofoam, or packing materials saved and collected from families or teachers in the school community helps to retain the heat of the sun.

WAVES AND THEIR APPLICATIONS IN TECHNOLOGIES FOR INFORMATION TRANSFER (PS4)

Nature offers so many opportunities to explore sound and vibrations. Students can listen to and identify different sounds in the environment. They can also investigate how sound travels differently through different materials, such as solids (striking a wooden object) and liquids (listening to underwater sounds). A tuning fork is an exciting way for students to see and feel sound vibrations. You can use a tuning fork in water in the classroom as well—but why not offer it outside? Watch the waves the tuning fork makes when it comes into contact with puddles. Have a local stream around? See if the same thing happens in a body of water that flows—why or why not? There are also many videos by musicians exploring water and sound. One favorite example is *Water Music* by Tan Dun and performed by Simantra Grupo de Percussão.

Making music instruments from found-nature objects is another great way to explore sound and vibrations. Shakers or rain sticks can be made with pebbles and seeds placed inside of leaves or cardboard paper-towel rolls. Flutes can be made from grass, straw, or reeds. Wind chimes can be created with pretty much anything tied to a stick (acorns, pebbles, pinecones, sticks, shells). Children can create drums out of anything—just give them a stick and permission to explore!

Using Chrome Music Lab is great for a unit on waves and vibrations. This can be a way for older students to explore manipulating sound waves on the computer first, and then presenting a challenge to see if they can apply the principles of sound that they learned from the Chrome Music Lab outdoors.

Earth Science

The best way for students to learn about the Earth is by going outside! No matter the location, they can investigate the availability and uses of water, soil, rocks, and minerals. If there is water near the school, students can see weathering and erosion firsthand. Stream banks are perfect for this. Prompts of "I notice/I wonder" are great ways to observe what exactly is happening between the land and the water. By examining the properties of different rocks, students can understand the rock cycle and the formation of Earth's crust. They can also explore the importance of conservation and sustainable practices in preserving natural resources. This encourages students to make the connection between a systems thinking approach and human impact.

Light and Shadows: Outdoor spaces are a great way to teach students about light and shadows. They can observe how shadows change throughout the day based on the sun's location and measure the length of shadows cast by different objects. Students can also investigate how sunlight is filtered through leaves, creating light and shadows on the ground.

Weather and Climate: Students can keep a weather journal and observe changes in temperature, wind patterns, cloud formations, and precipitation over time. They can measure rainfall using rain gauges or create wind vanes to understand wind direction. This is a great opportunity to incorporate Tinkercad and 3-D printing to design a rain gauge or an anemometer to measure rainfall or wind speed. By studying local weather patterns and comparing them to other regions, students can develop an understanding of climate and its impact on ecosystems.

Photo 7.3. Shadows in the Snow (Helen Corveleyn)

Teacher in the Limelight: Natasha Agrawal

Photo 7.4. Checking the Weather Station and Making Morning Announcement (Natasha Agrawal)

Natasha Agrawal is an English language teacher for multilingual learners at Robbins Elementary in Trenton, New Jersey. Over the past twenty years, she has taught all grades from preschool to grade five. In addition, she has taught English to the parents of some of her young students. Robbins Elementary is a K–3 urban school serving a mix of local African American and immigrant Spanish-speaking students from Central American countries. The students who participated in the weather program are in the Dual Language program, learning all content areas in both Spanish and English. Ms. Agrawal describes how she used a weather station with her students:

It was an exciting moment when Dr. Magee from The College of New Jersey came to Robbins Elementary to install our very own weather station in our schoolyard. A three-pronged device, it had a rain catcher with a measure, a thermometer, and a wind direction and speed measure too! Along with that came a console with all the numbers in different colors! What more could a third-grader want?

(continued)

This weather station was an exciting way to develop an interest in weather patterns in our young students. Each morning I coached two students to get the readings from the console, interpret them, and broadcast the weather report from the yard, standing right next to the weather station. It was a powerful moment for our students to be speaking on a video call to the entire school, talking about weather issues that affected outdoor playtime! After each announcement, we graphed the temperature and the rainfall. It was a visual for our students to understand the numbers in relation to hot or cold temperatures. Students developed vocabulary and used words like *flurries, fog, sprinkles, drizzle, precipitation*, in addition to describing the day as *sunny* or *rainy*. As we continued our announcements through the winter, we all hoped for that much-awaited snow day! But, alas, New Jersey had no snow last year, a point that most children noted and remembered and wrote on our weather charts. Definitely an indication of changing weather patterns.

My students developed speaking skills, weather awareness, observation skills, weather-related vocabulary as they made announcements each morning. Standing next to the weather station and announcing the weather report became the highlight of their day. Working with young, enthusiastic meteorologists each morning certainly became mine!

Photo 7.5. Learning Weather Vocabulary (Natasha Agrawal)

Math

The connection between mathematics and nature is profound and beautiful. Nature is filled with embedded patterns studied and appreciated by artists and scientists, including the golden ratio and Fibonacci sequence. Geometric patterns and shapes can be observed in flowers, honeycombs, snowflakes, stars, spider webs, and crystal formations. There are numerous opportunities to connect nature to math standards. The Common Core focuses on four domains in math in K–5: Counting and Cardinality; Operations and Algebraic Thinking; Numbers and Operations (in Base Ten and Fractions); Measurement and Data; and Geometry. We have selected just a few ways to teach math in each of these areas. Through these activities, students reinforce the practical application of math in the real world.

COUNTING AND CARDINALITY

For younger students, learning one-to-one correspondence and developing number sense are essential math skills that form the basis for addition and subtraction. Store-bought manipulatives and worksheets with ten frames are often used to develop these skills. However, ten frames and items can be made by natural materials—for example, sticks for the frame, and objects such as rocks or petals as markers or manipulatives. When outside, it is helpful to have chalk available for these types of activities because there are many teachable moments, particularly in kindergarten and first grade, as students begin to construct their own number sentences and engage in "math talk." Students in one class were so engaged in a ten-frame activity with snowballs that they excitedly decided to create a number sentence with the snowballs rather than in a ten frame because the snowballs were "real." They viewed the snowballs not as markers to substitute for numbers the way counters in the classroom would be but as actual numbers in a number sentence.

STEM Books for Teachers and Students

Teacher Books

- *Messy Maths: A Playful, Outdoor Approach for Early Years* by Juliet Robertson
- *Dirty Teaching: A Beginner's Guide to Learning Outdoors* by Juliet Robertson
- *Making and Tinkering with STEM: Solving Design Challenges with Young Children* by Cate Heroman
- *Awesome Engineering Activities for Kids: 50+ Exciting STEAM Projects to Design and Build* by Christina Schul
- *Teaching STEM Outdoors: Activities for Young Children* by Patty Born Selly

(continued)

> **Children's Books**
>
> - 40 Children's Books That Foster a Love of Math https://dreme.stanford.edu/news/40-childrens-books-that-foster-a-love-of-math/
> - *Math Curse* by Jon Scieszka and Lane Smith
> - *Math in Nature* series by Lizann Flatt
> - *Engineering Marvels Hoover Dam* by Nikole B. Bethea
> - *The Raft* by Jim LaMarche
> - *Lily and the Fairy House* by Jane Tanner
> - *The Wildwood Elves* by Anne-Marie Chapouton
> - *Panda's Magical Adventure* by Diane Ohanesian
> - *Symphony City* by Amy Martin
> - *Eyewitness Rocks and Minerals* (DK Eyewitness) by DK
> - *Rhoda's Rock Hunt* by Molly Beth Griffin and Jennifer A. Bell
> - *Ricky, the Rock That Couldn't Roll* (You Rock Group) by Jay Miletsky

OPERATIONS AND ALGEBRAIC THINKING

Searching for patterns: Students love searching for patterns around them in nature. These can be found in tree branches or flower petals. Patterns can also be found in sounds such as bird calls and water dripping.

Creating and extending patterns: Encourage students to create patterns out of natural materials and then give it a variable. This can start with extending a simple ABABAB or ABBABBA pattern and encouraging their own creations. They can also use sticks or other materials to play patterns on the ground, on trees, and in puddles or other bodies of water.

Nature stories: Have students create and solve word problems based in nature, such as: If a squirrel has four acorns and finds three more, how many does he have altogether? They can act out being the squirrel and use real acorns.

Number expressions: Engage students by "building" with sticks or any available objects (shells, seeds, acorns, pinecones) to help express their math thoughts. Put them in nature buddy pairs where one of them could build the number expression and the other could read it. This form of peer review is both validating and promotes self-awareness.

NUMBER AND OPERATIONS

Base 10: There is an abundance of natural items that can be used as a "set" of base ten blocks. Acorns and pebbles may pop into mind first, but small pinecones, dandelion heads, maple (helicopter) seeds could all work. Decide what is the best way to display this as a class: Put it on a poster? Put it into a clear box? Perhaps create a grid to make a more uniform display of the objects (if they are not all exactly the same—acorns would be extremely similar, but maple seeds may not).

Fractions: Natural materials are such an easy way to begin or enrich beginning fractions. Sticks, leaves, flowers are all FREE, and readily available. Any mess that is made can stay outside, so there is very little cleanup! Each student can see that a "whole" can be anything. A leaf, a long stem, a reed—whatever they choose. Creating two, four, and eight equal parts of that whole is a concrete way to show fractions.

MEASUREMENT AND DATA

Classifying found objects (leaves, sticks, or rocks): Natural objects can be categorized by type, size, color. Have kids decide which "go together" and why. Creating categories is also a great way to compare groups. Students can use chalk or Hula-Hoops outside or bring objects inside. They can develop comparison and measurement skills as they rank the leaves from smallest to largest. They can also put items in piles of a certain number.

Data collection over time: While this requires a significant commitment, and accountability for collecting on a daily or weekly basis, students can collect data on various elements in nature, such as weather patterns, plant growth, or animal behavior. They can then analyze the data they gather during their investigations by calculating averages, making predictions, and measuring and comparing different variables. They can interpret and represent their findings with graphs or charts. In the early years on data collection one of the best techniques is to have students try to make sense of the data first and then ask them to communicate their findings. If this feels too unstructured for students who need more concrete directions, create a checklist for data collection "sense-making." Some of the questions could look like the checklist in table 7.1.

Table 7.1. Checklist for Communicating My Data to Others

- Did I organize my data so that people are able to see what I did in the field?
- When I want to tell my audience what I found out, did I use a picture or a table to help explain it?
- Is there a way I can graph or show what happened over time?
- Will people be able to understand my explanation? Why? What will help them understand?
- Did I show any patterns or similarities that occurred in the data?
- Did something unusual or surprising happen in the data and did I show that to my audience?

CHAPTER 8

Rooted in Nature
AUTHENTIC FIELD METHODS

In a quiet forest, on a beautiful, crisp morning, birds are singing in the trees. The gentle crunch of leaf litter under hiking boots and sighs of wonder at nature's beauty echo up and down the trail. Scientists collaboratively test the pH of the soil, and leaves are monitored for caterpillar activity.

Sound delightful? This is a scene from a fifth-grade class trip. Not the typical sounds of a group of seventy-five 10-year-olds on a field trip to the zoo or aquarium? That is because authentic science shaped the behavior and outcome of the immersive field experience for students.

Unfortunately, many field trips to informal science environments still follow the "one-hit wonder" approach. Although teachers may include pre– and post–field trip activities, it may be limited to one or two short lessons. While this structure may work for certain learning outcomes, it is essential that field experiences in natural outdoor settings are integrated into the science curriculum. There is ample research on the benefits of outdoor spaces, including the schoolyard, for positively impacting science achievement (Rios & Brewer, 2014). Authentic science learning incorporates content, science processes, skills, and the tools of science (Fenichel & Schweingruber, 2010). Further, science practices in NGSS imply that we provide opportunities for young scientists to "practice" what they are learning from the traditional methods of teaching.

This chapter continues the ideas presented in earlier chapters to highlight ways to use authentic methods and experiences with students, including using nature journaling. We include lessons, activities, and a unit where students conduct authentic fieldwork as ornithologists, geologists, botanists, zoologists, and entomologists. Also included are descriptions of lessons from a fifth-grade ecosystems unit, where students increased their knowledge of science content, process, and tools through authentic field experiences. As students performed the jobs of specialized scientists, they learned how different disciplines work together to understand (and positively impact) the health of the environment.

Teachers: Invest in Yourself!

As an adult, doing your own fieldwork is a rejuvenating experience. After teaching for more than fifteen years, I (Helen) decided that I was ready to engage and develop my scientific side a bit more deeply and I decided to pursue my master's degree in conservation education. As I've always been a huge proponent of fieldwork, I looked for a program that would support my own learning philosophy. Project Dragonfly, out of Miami University of Ohio, has a program that offered me the opportunity to study in Baja, Borneo, and Belize. The experiences in the field caused me to be introspective, allowing me to reexamine how I was teaching students biological concepts and make sure that academic rigor came in the form of real-world science experiences.

During this graduate program, I was challenged to design my own scientific inquiries using tools specific to the ecosystem we were studying. When I used an underwater slate to create a species count of sea stars while snorkeling, I felt like I had truly made it in my career! I also had the opportunity to track the highly elusive pangolin in the forests of Borneo using a radio transmitter. These experiences were life changing and I knew I needed to bring this feeling of scientific triumph home to my students!

Fieldwork is physically demanding; it makes you think on your feet if the methods and protocols you've settled on don't work; it often stretches you to your limits but makes you incredibly resilient and allows you to feel accomplished in a way that classroom learning often does not. These experiences are what we need to offer students to provide rigor, perseverance, and grit in a way that is appealing to their sense of adventure and discovery.

There are many opportunities for teacher fieldwork in the form of summer professional development with the National Park Service, conservation associations, governmental entities, local watershed opportunities, coastal marine organizations, and so on. These opportunities can help to stretch your thinking, provide confidence in the science and educational concepts you are bringing back to your students, and reignite your own learning as an adult.

Photo 8.1. Helen Conducting Research in Belize (Helen Corveleyn)

Nature Journaling

Nature has always been a source of inspiration for artists, poets, and inventors. In nature, science and art intersect; the separation of art and science makes little sense. Historically, the greatest scientists documented the natural world through artistic renderings and great artists were scientific in their approach. For example, Galileo Galilei created drawings of his observations of the phases of the moon. Through creating these drawings he concluded that the moon was not a perfect sphere, but had mountains and valleys. He also documented the movement of sunspots. He drew the spots at the same time each day and was able to record the movement of the spots. Leonardo da Vinci seamlessly integrated art and science in his research and drew everything from the flight of birds to the motion of water. Claude Monet was fascinated with light and painted the same site again and again, recording how its appearance changed with the time of day. Albert Einstein wrote that "art and science are the branches of the same tree."

To focus on nature as a teaching tool, students must be able to play, enjoy, and notice. Using art to "document" nature encourages students to be careful observers. The focus is in process rather than product, as students observe, label, and question as they sketch. There are incredible resources available for teachers and students for incorporating nature journaling in the curriculum. John Muir Laws has many books and webinars, as does Clare Walker Leslie with activities for every age group.

Nature journaling can be used across the curriculum areas. In addition to developing cognitive skills such as close observation, asking questions, and learning scientific language, nature journaling also supports social-emotional development. It can be a very calming and mindful experience to work on a nature journal; it promotes self-awareness and a deeper understanding and connection to nature.

A favorite journaling introductory activity recommended by John Muir Laws and others is blind contour line drawings. Students find a quiet outdoor spot to sit with a blank piece of paper and pencil. They are charged with drawing something right in front of them for a minute without looking at the paper and not lifting their pencil off the paper. It should be one continuous line. From there, teachers can instruct students to look and draw birds, plants, and insects in detail. This activity allows for close observation while leveling the playing field for "artists" and "non-artists."

Another nature journaling activity that is a favorite is sound mapping. Sound mapping involves sitting in a spot outside and noticing all of the surrounding sounds on a map. The student can draw themselves in the center, and then whenever they observe a sound, they note that on the map. This is an engaging activity that goes beyond visual observations to develop their sense of hearing as they explore the natural environment.

> ### Resources for Authentic Field Experiences
>
> **Books**
>
> - *Keeping a Nature Journal: Deepen Your Connection with the Natural World All around You* by Clare Walker Leslie
> - *How to Teach Nature Journaling* by John Muir Laws
> - *Exploring Soils: A Hidden World Underground* by Samantha Grover
> - *Kaufman Field Guide to Insects of North America* by Eric R. Eaton and Kenn Kaufman
> - *A Field Guide to Eastern Forests, North America* by John Kricher and Gordon Morrison
> - *National Geographic Kids Bird Guide of North America* by Jonathan Alderfer
>
> **Websites**
>
> - John Muir Laws, Nature Stewardship through Science, Education and Art https://johnmuirlaws.com/
> - Bio-Regional Field Guides for North America. Bio-Regional Field Guides for North America | Wilderness Awareness School https://www.wildernessawareness.org/content/bio-regional-field-guides-north-america#SW
> - Is it endangered? The IUCN Red List of Threatened Species https://www.iucnredlist.org/
> - National Invasive Species Information Center (NISIC). National Invasive Species Information Center | United States Department of Agriculture, n.d. https://www.invasivespeciesinfo.gov/
> - Producers, Consumers and Decomposers Game. Kid's Corner. Sheppards Software https://www.sheppardsoftware.com/content/animals/kidscorner/games/producersconsumersgame.htm
> - Why Would a Hawk Move to New York City? Mystery Science https://mysteryscience.com/ecosystems/mystery-1/food-chains-predators-herbivores-carnivores/119?r=5518597
> - Wolves of Yellowstone, National Geographic Society https://www.nationalgeographic.org/media/wolves-yellowstone/

Tools and Resources for Budding Naturalists

As part of the science curriculum, it is essential that students learn to use authentic scientific tools. Some of the tools are used only by scientists, but other tools are accessible and used by scientists and nonscientists alike.

Binoculars: Although binoculars are more common these days, it is important to note that many students have not used them before. Explicit directions on how to use binoculars may include how to steady them by leaning your arms on your body or a nearby fence, how to focus, which way to look through the viewfinder, caring for the lenses, and remembering to keep the strap around your neck. Even if the binoculars are "toy" models, showing young children care and respect for a tool builds scientific habits for the future. Teaching these skills gives students a sense of responsibility and supports student success. While it seems obvious at first, when the teacher takes their time to explain and demonstrate how a seemingly easy tool is used in the field, typically students will pay better attention and treat the tool with more care and respect than they would without direct instruction.

Field guides, posters, and handouts: A class library with field guides is a wonderful and authentic way to link nonfiction texts to the science curriculum. Books and posters showing trees, plants, wildflowers, scat, and animals such as birds are great resources. Pictures from older field guides (these can be found at library sales) and magazines can be used by students to make their own field guides. However, it is important to not overwhelm young students with identification. It can be tedious and some students will lose focus if overloaded with information. We recommend focusing on just a few birds, plants, or trees that the students will definitely encounter. Having a scavenger hunt or handout of just a few things to look for will motivate kids to pay attention and really look around.

Apps such as iNaturalist, Seek, Picture This, and eBird (Cornell): These apps can deepen interest and elevate a lesson both for identification and for the thrill of connecting with other scientists and having the feeling of doing "real science." This works best with older kids; or if you have a young group that is particularly interested, the adults you bring outside with them can handle the technology as a side-by-side activity with the adult.

Design a bird/squirrel blind: Transform a simple sheet or a large piece of butcher-block paper into an interactive bird/squirrel blind by hanging it on a chain-link fence in proximity to an area where birds and squirrels reside. Cut eye holes at the height of your students, and approximately an hour before their visit, scatter seeds on the ground to attract birds and squirrels. Engage the students by explaining how they are "camouflaging" themselves to minimize any disturbance and emphasize that they are acting as real scientists during this observation. This immersive experience provides a unique opportunity for students to witness avian and animal activity, especially during the lively mornings of spring.

Literacy Extension: *Actual Size*

At HES we had a wonderful transdisciplinary project that was conceived by our school librarian based on the book *Actual Size* by Steve Jenkins. During this extended project, we took a deep dive into the book, examining the way the author portrayed scale for his readers. We then brought students into the field—a stream ecosystem behind our school—and challenged them to choose something from nature that they could identify, measure, research, and write a page on in the style of Steve Jenkins. Equipped with magnifying glasses, clipboards, scaffolded sheets for identification, field guides, and, of course, rain boots, we excitedly burst into the field. Biotic and abiotic choices included: rocks, flowers, tadpoles, mayflies, worms, wildflowers, trees, fungus, slime molds, algae, and mosses.

Fifth-Grade Ecosystems Unit

This fifth-grade ecosystems unit immerses students in authentic science learning in multiple contexts: the classroom, schoolyard, and a county park. As indicated in table 8.1, students experienced a continuum of "wilderness" as the unit progressed from artificial to more natural learning spaces.

Table 8.1. Overview of the Ecosystems Unit

Place (Ranges from artificial to increasingly natural/wild spaces)	Activities	Time (Approximate)
Classroom	Beginning of unit • Phenomena: Discuss photos of healthy and unhealthy ecosystems and create classroom anchor charts to reference throughout the unit • Review safety protocols	Week 1
	During unit • Online research • Choice boards • Ecosystem health simulation games • Expert groups develop protocols and generate graphic organizers • Mixed expert project teams each create a plan for holistic assessment of the ecological health of the forest during the hike • Review safety procedures	Ongoing
	End of unit • Students collaboratively complete the Ecosystem Health Report and share results in Google slideshow presentation	Week 5
	After unit • ELA teachers use unit as basis for informational text writing unit	After unit
Schoolyard excursions	• School garden: Quiet observations and "I see, I wonder" sentence strips	Week 1–2

Place (Ranges from artificial to increasingly natural/wild spaces)	Activities	Time (Approximate)
	• Outdoor classroom: Share findings and develop four fields of study framework • Field tools introduction	
The "in-between": Walks from one place to another	• Connections to previous learning • Spontaneous discovery	Ongoing
School/wilderness edge (natural/unmowed areas of school property or park)	• In a series of outdoor lessons, students learn how to use tools from each discipline (geology, botany, zoology, and entomology).	Week 3–4
Hiking field trip	• Project teams survey ecosystem health and compare their data to ecosystems they have visited prior to the hike. Students use their surveying tools and data recording templates	Week 5

In the unit, students explore ecosystems with authentic tools used by geologists, botanists, zoologists, and entomologists. Table 8.2 shows the different activities and their alignment to the NGSS and the Common Core Standards. Through investigating, questioning, researching, and communicating their findings in various settings, students are able to experience real-time science in action. We describe activities Helen led as part of the unit, however, these activities can easily be adapted to standalone lessons at different grade levels.

Table 8.2. Ecosystems Unit Alignment with Fifth-Grade Standards

NGSS Disciplinary Core Idea	Classroom Activity
ESS3.C: Human Impacts on Earth Systems Human activities in agriculture, industry, and everyday life have had major effects on the land, vegetation, streams, ocean, air, and even outer space. But individuals and communities are doing things to help protect Earth's resources and environments. (5-ESS3-1)	Students found evidence in nature of the human activities that add to and prevent ecosystems from being healthy and balanced. Students researched how humans impact the health of an ecosystem through behaviors and intervention. Students observed and appreciated the value of nature while conducting field investigations, thus leading to an increased appreciation for nature and a desire to want to protect it.

(continued)

Table 8.2. *Continued*

NGSS Science and Engineering Practice	*Classroom Activity*
Obtaining, Evaluating, and Communicating Information Combine information in written text with that contained in corresponding tables, diagrams, and/or charts. Use multiple sources to generate and communicate scientific and/or technical information orally and/or in written formats, including various forms of media and may include tables, diagrams, and charts.	Students created data tables and sampling protocols to collect information in a variety of settings. Students made general observations of ecological health as well as plant, soil, animal, and insect observations. Students developed ecosystem health reports based on the data collected, and shared the reports with their peers.
Planning and Carrying Out Investigations Evaluate appropriate methods and tools for collecting data. Make observations and/or measurements, collect appropriate data, and identify patterns that provide evidence for an explanation of a phenomenon or test a design solution.	Students determined tools they will use to inform other scientists in their group how they will assess ecosystem health in a variety of locations. They developed consistent protocols to carry through on an investigation.
Asking Questions and Defining Problems Identify scientific (testable) and nonscientific (non-testable) questions. Ask questions based on careful observations of phenomena and information. Ask questions to clarify ideas or request evidence.	Students created and shared their questions from the school garden ecosystem, which were sorted into soil, animal, plant, and insect categories. Students determined if their questions were "Google-able" or required observation and testing. Students determined if they needed more information based on questions.
Analyzing and Interpreting Data Display data in tables and graphs, using digital tools when feasible, to reveal patterns that indicate relationships. Use data to evaluate claims about cause and effect. Compare data collected by different groups in order to discuss similarities and differences in their findings.	Data was recorded, assessed, and communicated to others in the group, and to the greater class of scientists as a whole. Students could choose to create graphs, videos, or charts to explain relationships within the ecosystem. Students created an ecosystem health report based on data collected on a Google slideshow.

NGSS Crosscutting Concept	Classroom Activity
Systems and System Models Students understand that a system is a group of related parts that make up a whole and can carry out functions its individual parts cannot. They can also describe a system in terms of its components and their interactions.	Students observed and made connections on the interrelatedness of earth spheres and ecological functioning. Students observed, classified, and identified how separate aspects of the ecosystem fit together as a whole to create harmonious balance.
Common Core State Standards (ELA)	*Classroom Activity*
RI.5.9 Integrate information from several texts on the same topic in order to write or speak about the subject knowledgeably.	Students used multiple websites, printed articles, and reference books to conduct their research and develop their protocols.
W.5.8 Recall relevant information from experiences or gather relevant information from print and digital sources; summarize or paraphrase information in notes and finished work, and provide a list of sources.	Students developed connections between what was observed in the natural world and information retrieved from current scientific text on topics like invasive species, biodiversity, and energy transfer.
W.5.9 Draw evidence from literary or informational texts to support analysis, reflection, and research.	Students conducted research and reflected on the information collected in Google Keep notes and through final written products. Students were required to have multiple text references to inform their readers about threats to the ecosystem.
	Students used Google Sites for informational writing to inform others of the human impact we have on ecosystems. Students in other classrooms viewed the website.
Common Core State Standards (Math)	*Classroom Activity*
5.MD.A.1 Convert among different-sized standard measurement units within a given measurement system (e.g., convert 5 cm to 0.05 m) and use these conversions in solving multi-step, real-world problems.	Students used a quadrat to measure and estimate plant density. The quadrat was one meter squared and students could multiply a sample size to estimate plant density in a larger area of forest.

START WITH PHENOMENA

We began the unit in the classroom with a startling set of pictures: the same ecosystem, five years apart. The first photograph reveals an inviting forest: mature trees, green ferns set against a backdrop of brown leaf litter with a fog of mist settling on a quiet fall morning. The prompt asked students to provide descriptions of the forest. They responded with statements such as, *"It looks like an area where you would feel calm walking through it"* and *"The forest looks like it's been around for a while; we can't tell what season it is except that there are a lot of leaves on the ground."* Students then viewed the second photo of the same forest five years later. It is very different in the subsequent picture: covered with bright-green plants with white flowers providing a sense of lush ground cover throughout the intermittent trees. Student responses, written on an anchor chart, revealed a mixture of misconceptions, questions, and observations. In viewing the second picture, students exclaimed, *"Wow! That forest really changed! It is so healthy looking. It looks like there is green for miles!"* and *"This forest is absolutely springtime! You can tell by how productive the ground is . . . alive with plants!"* The students were observing evidence of a highly invasive plant: garlic mustard.

It was the first picture that showed a healthy productive forest with a high biodiversity of plant life, evidence of decomposers in the leaf detritus on the ground, and a multitude of different tree species. The juxtaposition with the second photograph allowed for rich discussion with and among the students, bringing to the surface the common misconceptions of "green" and "lush" as indicative of ecosystem health. The discrepant photos led to exciting discussions and created a strong foundation for the unit.

DISCOVERING FOUR FIELDS OF STUDY THROUGH OBSERVATIONS AND QUESTIONS

We followed the introductory phenomena with a visit to the school garden. In early fall the garden is in all stages of development, ripe for micro-ecosystem study. We gave the students "I see/I wonder" science sentence stems to guide their individual observations. We told the students to be "quiet observers" so as not to influence other students' thinking and not disrupt the ecosystem inhabitants. In addition to writing their observations, students jotted down questions that came to them while they observed.

Next, we met in the outdoor classroom and students reported their observations to the group and in their science notebooks. They shared their questions, which we sorted into soil, animal, plant, and insect categories. Through this process of *Asking questions and defining problems* (NGSS), four fields of study emerged, providing a framework for the rest of the unit: geology, botany, zoology, and entomology.

Table 8.3. General Schoolyard Observation

Name: _____

GENERAL
Ecosystem Health Investigation

Target Question: How healthy is the ecosystem that is surrounding our school?

General Evidence: Collect evidence from a small sample of our local ecosystem.

Site Description and Sketch:

Prediction: _____

If you had to name this ecosystem, what would you call it? *circle the best response; you can circle more than one	Meadow/Field	Developed	Forest
	Yes		No
Water Present (check one)			
Animals/evidence of animals (check one)			
Wind (check one)			
Time of day	Morning:		Afternoon:
Temp	Celsius:		Fahrenheit:

> ### Place-Based Authentic Learning
>
> It is important to note that the short walks to and from outdoor destinations were not interludes from learning, but an integral part of the field experience. Many spontaneous scientific discoveries occurred on walks "between" spaces. Further, these walks were opportunities to revisit concepts from previous lessons taught indoors. For example, students had a previous classroom lesson on the spheres of the Earth (atmosphere, geosphere, hydrosphere, and biosphere) aligning to fifth-grade Next Generation Science Standard 5-ESS2.1. In the classroom, we played the game four corners and correctly identified every single description. During the walk outside, the students were asked to explore the geosphere. The expectation was that they would dig in the soil. Looking confused, one student asked, "Geosphere? That's not out here where we are." This illustrates that while the indoor classroom provides a space to write, think, and process scientific concepts, students need opportunities to transfer and apply their learning in authentic natural spaces.

USING AUTHENTIC FIELD TOOLS

The next phase of the unit involved numerous visits to more natural (unmowed) areas at the edge of the schoolyard. Several lessons introduced students to the actual field tools that each discipline uses to measure ecosystem health. As *botanists*, they used a teacher-prepared quadrat made from meter sticks and string to sample plant biodiversity and plant population density. This offered an excellent opportunity to talk about mathematical implications using each square of the quadrat to multiply the density of the species variety of the plants. As *geologists*, they used soil pH meters, soil core samplers, moisture, and fertility meters. Students explored the chemical composition of the soil and researched what would cause changes in soil composition. As *zoologists* they observed evidence of animals: nests, scat, and burrows. This challenged their powers of observation and deductive reasoning to track evidence of animals that they could not physically see. Finally, as *entomologists*, they searched for insect egg clutches, webs, and insects. Students identified insects, using the iNaturalist and Seek apps to help in this process. Students had graphic organizers in their notebooks (see tables 8.4–8.7) and access to field guides, magnifying glasses, rulers, hand trowels, and iNaturalist on teacher handheld devices.

Table 8.4. Schoolyard Observation—Plants

Name: _____

PLANTS
Ecosystem Health Investigation

Sphere: _____

Target Question: How healthy is the ecosystem that is surrounding our school?

Plant Evidence: Collect evidence from a small sample of our local ecosystem.

Evidence Questions	Factor: Plant Observations — Evidence
Describe how much sunlight the ecosystem is getting currently. *Circle your response	Time of day: _____ Date: _____ Describe the weather: _____ Full Sun Partial Sun (shaded) No Sun
Approximately how many different types of plants do you see in your sample?	
Is there one plant that is overrepresented in your sample (i.e., there are many more of this one type of plant than the others)?	Yes No If yes, what plant? _____ Is this an invasive species? _____
List some of the plants that you see. *Use the identification chart	_____ _____ _____ _____
Questions:	• _____ • _____

Table 8.5 Schoolyard Observation—Soil

Name: _____

SOIL
Ecosystem Health Investigation

Sphere: _____

Target Question: How healthy is the ecosystem that is surrounding our school?

Evidence: Collect evidence from a small sample of our local ecosystem.

Evidence Questions	Factor: Soil Observations	
	Evidence	
What color is the topsoil in your sample section? *Choose a colored pencil and record a thick line.		
Is over 50% of the soil covered by leaves and/or plants?	Yes	No
Use the 3-Way Digital Analyzer to get the Fertility, pH, and Temperature of the Soil.	Fertility: _____ pH: _____ Temperature: _____ **Fertility:** *Ideal rating is between 3 and 7 (combination of Nitrogen, Phosphorus, and Potassium)* **pH:** *Plants vary on pH needs* **Temperature:** *Plant germination and growth temperature requirements vary*	
Use the Light and Moisture meter and record the letter (light) and number (moisture).	Light (should be a letter): _____ Moisture (should be a number): _____ *See chart for light and moisture needs of plants*	
General Useful Observations/Questions: *You may use the core sampling tool to gather this evidence. **You may sketch what you see, use the colored pencils.	• _____ • _____ Sketch	

Table 8.6. Schoolyard Observation—Animals

Name: _____

ANIMALS
Ecosystem Health Investigation

Sphere: _____

Target Question: How healthy is the ecosystem that is surrounding our school?

Evidence: Collect evidence from a small sample of our local ecosystem.

Evidence Questions	Factor: Animal Observations		
	Evidence		
How many different animals can you see in your sample area? *You may use evidence that an animal was present.			
Do you see any scat (poop)?		Yes	No
Do you see scat from different animals?		Yes	No
Do you see evidence that animals have been eating any of the plants? *Look for holes or bites in leaves.			
How many different types of animals did you see on the wildlife camera?			
Questions:	• •		

Table 8.7. Schoolyard Observation—Insects

Name: _____

Insects
Ecosystem Health Investigation

Sphere: _____

Target Question: How healthy is the ecosystem that is surrounding our school?

Evidence: Collect evidence from a small sample of our local ecosystem.

Factor: Insect Observations		
Evidence Questions		*Evidence*
How many different insects can you see in your sample area? *You may use evidence that an insect was present.*		
Do you see any webs or insect burrows?	Yes	No
Do you see evidence that animals have been eating any of the plants? *Look for holes or bites in leaves.*		
How many different types of insects did you see/have evidence of?		
Questions:	• _____ • _____	

ECOSYSTEMS ONLINE RESEARCH

After our outdoor excursions on school property we returned to our big question, "What is ecosystem health and how do we recognize it?" Armed with new observations and a lot of questions, students began their research on ecosystems. We provided students with flipped learning lessons and choice boards to investigate ecosystems around the United States that were unhealthy and conservation efforts to restore the natural balance of the ecosystem (5-ESS3-1). National Geographic's webpage on the wolves in Yellowstone National Park is a wonderful resource illustrating the restoration of a large ecosystem after apex predators were reintroduced to help balance the ecosystem. Students looked at online, web-based modeling that demonstrated fluctuations in food webs, energy pyramids, and ecosystem health. Students also played simulation games (5-LS2-1) to experience large-scale systems' interactions. In addition, they accessed local community conservation group websites that outlined the invasive species in the immediate area. This was an excellent way to research local threats to the ecosystem, as conservation organizations and government websites are current and updated on invasive species and proper mitigation techniques.

THINKING AND LEARNING LIKE A SCIENTIST

After conducting research and experimenting with the tools outside, each student self-selected a science role (geologist, botanist, zoologist, or entomologist). They formed "expert" groups in each of the four areas and became proficient with the tools needed for outdoor fieldwork. This allowed them to "try on" a STEM career and create a scientific community that mimics what happens in reality. It also gave them a sense of ownership of their learning and a deeper connection to the science discipline that they chose. In their specialized groups, they developed protocols and generated graphic organizers for their work on the culminating field trip to the forest.

In addition, they jigsawed into project teams with one representative from each discipline. Their common goal was to combine their knowledge to create a plan for how they would determine the holistic ecological health of the forest ecosystem they would visit for their field trip. They each had their own area of expertise within their group, enforcing the practice of communication and collaboration.

PREPARING FOR THE FIELD TRIP

The culminating field experience for the unit was a county park. The park has about twelve miles of wooded hiking trails and the highest mountain peak in our area. While this trip took place in a deciduous forest in the northeast, the unit can be adapted to any ecosystem: the desert, coastal lands, prairie, or wetlands. To prepare for the trip, we discussed safety, clothing guidelines in consideration for ticks and weather comfort, and the ethics of interacting with nature and leaving no trace on the land. Students carried their lunch, water bottle, a clipboard, and the tools they needed for the field study.

Table 8.8. Determining Useful Evidence

Name: _____

Ecosystem Health Investigation

Target Question: How healthy is the ecosystem that is surrounding our school?

Useful Evidence: Record the most useful evidence from each of the factors that we studied at our site.

Soil	
Plants	
Animals	
Insects	

Claim: Use the evidence to record your claim.

Claim: I think _____

Do not explain why in your claim, just answer the question!

Reasoning: Explain how your evidence supports your claim (you may use multiple pieces of evidence).

_____
_____
_____

Table 8.9. Student-generated graphic organizer for soil study

Name: _____ Date: _____ Period: _____

Soil

Topsoil color: _____

Is over 50% of the soil covered by plants?   Yes   No

Temperature: ___
Moisture: ___
Humidity: ___
Sphere: ___
Fertility: ___
Light: ___
pH: ___

Observations and/or questions:
Use sketches, samples, and anything you need in order to record interesting facts you see on the trip.

Write down what you notice.
This may be any evidence for your claim.

_____
_____
_____
_____
_____
_____
_____
_____
_____
_____

Sketch or take some samples of anything you think may be interesting or helpful for your claim.

For your claim, answer this question: From the evidence you gathered do you think this ecosystem is healthy?

_____
_____
_____

We knew students had varying physical abilities and hiking experience, and that for some it would be a new challenge not to have bathrooms immediately available. We planned for necessary accommodations, and discussed with students what to expect ahead of time. We split the fifth graders into two smaller groups so as not to crowd the trail. One group hiked up the trail first and sampled at the end, while the other sampled first and then continued on to meet the other group at the end. In the case of students who were unable to access the entire trail, we made accommodations in order for them to join the group who sampled first in an accessible forest site, and then to use an ADA-equipped trail to meet students at the end for the picnic.

HAPPY HIKERS: EXPLORING A FOREST AS A SCIENTIST

Taking a hike as a scientist is different from hiking to appreciate the beauty of nature. Our students had purpose. In their project teams, they surveyed the trail for evidence of biodiversity, climate indicators (lichens), the presence or absence of invasive species, and evidence of animal life. Each scientist had their specific protocol for the data they were gathering. For example, "botanists" looked for the invasive species they researched. The "geologists" compared soil pH to the soils they previously tested, while the "entomologists" noticed that the insect species were similar to those species just ten miles away at school. They approached the trip with more than casual interest or appreciation of leaving school for the day. They exhibited a sense of ownership over their learning, their role as a scientist, and their role in nature. We hiked up natural rock steps with whispers of *"Isn't this the greatest hike you've ever done?"* and *"Wow, these rocks look dangerous. Don't you love danger?"* As teachers, we proudly watched 10-year-olds function as an interdisciplinary group, sharing their findings, explaining their "specialty area" to each other, and engaging in scientific discussions.

The students' enthusiasm continued on the bus ride home, with conversations that included, *"Don't you wish we could do that EVERY week?"* and *"Wow, I wish we could do that hike in each season. It would be really cool to see how the ecosystem changes at different times of the year."* Many reflected that they would love to hike more often and lamented that there was just not enough time to get their work done. One student said, *"I noticed more nature because I had the tools to explore."*

POST-TRIP ECOSYSTEM REPORT

In the days following the trip, each group of student scientists (geologists, botanists, zoologists, and entomologists) collaboratively created an ecosystem health report that they shared with other groups (table 8.9). These reports required that students use the data from each area they sampled to make an evidence-based claim about ecosystem health. When student questions emerged as they analyzed their results, they accessed the selected websites for further research. They used their scientific communication skills to report out their findings within their mixed groups and with their discipline-based groups. Within those reports, they made comments such as, *"While the soil showed some promise in fertility, the lack of moisture is a concern."* They presented their slideshows to classmates to provide an analysis of the health of the ecosystems they experienced.

Photo 8.2. Testing Soil (Helen Corveleyn)

NOT JUST SCIENCE! ELA CONNECTIONS

Following the ecosystems unit, the ELA teachers used the unit as a foundation for their informational text writing unit (CCSS.ELA-Literacy.W.5.8 and W.5.9). Students brainstormed "threats to the ecosystem" and each student chose an environmental or ecological issue to research. The research guidelines required students to inform and educate readers about threats to the ecosystem and create a call to action to support ecosystem health. They created a website using the Google Workplace (Google Sites) to communicate their ideas with a wider audience. In a shared writing celebration, students referred to their ecosystem study as an inspiration for their Google Sites. Their writing was compelling, as they had a personal stake in assessing ecosystem health.

Setting Behavioral Expectations for Successful Outdoor Experiences

We have included three recommended lists for ensuring safe participation outside and on field trips. These can be used by teachers as verbal expectations, posted expectations, or through a written contract with students in field notebooks.

Behavior Expectations for Working Outside

1. We are outside to study and participate! Careful observation is required at all times so that you don't miss anything! It is not a time for running, playing, or extra socializing.
2. Please do not eat anything while we are in the field.
3. When you choose a study site, be sure to be in clear sight of at least one teacher. You must also be within earshot of the teacher whistle.
4. Use all tools with respect, and take care not to break or misuse the equipment. Tools should be used for only their intended purpose.
5. Stay with your group members. You should be supporting them as scientists, offering positive feedback, and troubleshooting when you are unsure of a protocol.

Conservation Expectations to Ensure Plant and Animal Safety

1. When observing animals and insects, remember you are observing them in the wild. No matter how small they are, for this study we are merely looking at the organism. At no time should you interact with the organism.
2. Create minimal impact on your ecosystem. If you look back to your study site and there is very little to no trace that you were there, BRAVO! That is what we are trying to accomplish.
3. If there is a need to move or change any part of the ecosystem, your group must set it back the way it was found. Tread lightly.

Safety for Hiking

1. While on the trail, all students must wear proper attire including: rubber-soled, supportive footwear such as sneakers, hiking boots, or walking shoes. No flip-flops or sandals will be permitted. Students should be encouraged to wear hats for sun protection and protection against ticks.
2. Students and teachers should bring water to stay hydrated.
3. Students must stay on the designated trails with an adult and their group.
4. In order to prevent contact with potentially poisonous plant life, students must be made aware of the flora.
5. Unidentified flora and fauna should not have direct contact with skin.
6. All students must check for ticks and other harmful insects before leaving the wooded area or upon arrival at school.
7. Teachers should carry a first aid kit for minor injuries. In addition to the standard first aid kit items, it is helpful to include: Afterbite, ice pack, extra trash bags, gloves, a compass, sunscreen, epi-pen, and medications specific to student need. In case eyewear is needed for digging in dirt, the teacher should bring a few pairs of safety goggles.

CHAPTER 9

Learning from the Past, Changing the Future
SOCIAL STUDIES, ADVOCACY, AND ACTION

The National Council for the Social Studies (NCSS) emphasizes that in elementary school, social studies (including history, geography, economics, and government/civics) is interdisciplinary and integrates language arts, visual and performing arts, and STEM. Students can learn about the past while exploring environmental issues such as pollution, deforestation, climate change, and environmental sustainability. Through hands-on experiences in nature, students develop a strong connection with the environment and gain a deeper understanding of the interdependence of natural and social systems. They can engage in meaningful projects that promote conservation, restoration, and sustainable practices both locally and globally.

Learning History through Agriculture and Gardens

As a science teacher tasked with teaching social studies in fifth grade, I (Helen) struggled to teach historical concepts with fidelity, as I constantly approached it through a STEM lens. After a while, I stopped fighting it. I thought to myself, why not just teach history and add the natural world into the storyline? I began to look for a variety of different activities that supplemented the topics I was teaching that were STEM related. Many of them centered around agriculture and gardening, and students loved them. The Aztec STEM challenge was a favorite!

AZTEC FLOATING ISLANDS STEM CHALLENGE

This is a great lesson that focuses on the Aztecs' challenge of creating fertile lands in a swampy area. With no traditional arable lands available, the Aztec people had to find a solution to feed their population using a lake as a resource. Their answer was to build a series of islands that floated in the lake but were anchored to the bottom, forming a system of canals for agriculture. When teaching this, it can be helpful to

Photo 9.1. Aztec Floating Island Challenge (Helen Corveleyn)

show a short video clip explaining how the *chinampas* worked (see the video clip in Teacher Resources at the end of the chapter). This discussion presents an opportunity to align with science standards and concepts, such as Earth systems, utilizing multiple solutions to solve a problem, and the process of formulating and iterating on ideas before arriving at a final solution.

The problem can be framed to students as: "Marshy areas in the Aztec landscape prevented the creation of good, fertile soil for food production." The challenge is for students to design gardens that can survive in an area mostly covered in water. Constraints for the activity can include:

- Gardens must be anchored to the bottom in some way.
- Gardens must be able to hold soil and grow seeds.
- Gardens cannot exceed a size of 10 cm × 10 cm.
- Gardens must be accessible from the side.

Materials for students can include natural materials such as reeds, sticks, and vines, or popsicle sticks, twine, thin cotton string (offering different weights and thickness to simulate different natural reeds, vines, and ropes), natural-colored clay, and hydroponic oasis cubes to hold the seeds. The choice of seeds is also crucial for this project, so students will need to conduct preliminary research to identify suitable options that germinate quickly (wheat grass works well). Each group should have access to three corks (these can be upcycled or purchased beverage corks) for their garden design. The lesson also requires a container that can hold up to four inches of water. These can be individual tubs or one large pool (like a baby pool). While it is a bit of work to fill the water of the baby pool model, it is worth it for the joy and excitement it brings into the classroom. By engaging in this hands-on activity, students can gain a deeper understanding of the engineering challenges the Aztecs faced and the innovative solutions they devised. They will develop critical thinking skills, problem-solving abilities, and an appreciation for the connection between nature and human innovation throughout history.

TEACHING US HISTORY THROUGH AGRICULTURE AND GARDENS

Gardens can be planted as a way to study different eras. For example, Kidgardening.com has lessons for Victory gardens connected to studies of World War I and II. Students can explore the agricultural practices of Native Americans and their profound influence on the colonists. For example, they can plant a Native American garden, such as the three sisters garden (corn, climbing beans, and squash) or a colonial garden with vegetables, herbs, and/or flowers that settlers would have planted. This can lead to meaningful discussions on how Native American knowledge impacted crops and the farming practices of the colonists, and how colonist farming and deforestation practices significantly shaped the landscape of early America. Students can also research the advancements in agricultural technology and their impact on food production, sustainability, and global trade.

NATIVE PLANTS AND THE REVOLUTIONARY WAR

Cultivators of the earth are the most valuable citizens. They are the most vigorous, the most independent, the most virtuous, and they are tied to their country and wedded to its liberty and interests by the most lasting bands.

—Thomas Jefferson

During the summer of 1776, the act of planting and gardening held deep political significance, symbolizing domestic self-sufficiency and liberation from English rule and taxation. General George Washington, while preparing his troops for battle in New York, emphasized the importance of creating gardens filled with American native plants. At his Mount Vernon plantation in Virginia, he specifically prohibited his plantation manager from cultivating English trees, thus showcasing his belief in the symbolic power of native plants. For Washington, trees represented America's strength and the abundance of its natural resources. As such, he ensured that the ships defending Boston displayed flags portraying green pine trees, further emphasizing the connection between trees and the country's prosperity.

This compelling narrative about Washington's dedication to native plants can serve as an authentic and engaging way to integrate both science and social studies into the curriculum. It is a great starting point to discuss the ecological importance of preserving native flora, promoting biodiversity, and maintaining a sense of national identity through the stewardship of native species. Students can research and make comparisons of the landscape prior to colonization and today, and take part in planting native plants.

THREADS OF HISTORY: THE USE OF FIBERS ACROSS TIME AND PLACE

Throughout history and across cultures, humans have used fiber to craft essential items for all aspects of life, from housing and clothing to toys, blankets, upholstery, furniture, and books. This universal reliance on fiber presents a meaningful way to learn about other cultures, past and present. Through the study of fiber, students can discover the traditional sources of various fibers and the impact of trade and globalization on the fiber industry. This exploration can include discussions on fair trade and sustainable practices in fiber production, fostering an understanding for the need for responsible industry practices. There are also many connections to be made between fiber and STEM, specifically in design and technology. Students can create projects that incorporate fibers, such as designing clothing, or accessories.

Weaving

Weaving is a wonderful way to connect to history. At HES, the art teacher collaborated with the third-grade teachers' social studies unit with a lesson on the many ways Native Americans and early settlers weaved items: for decorative purposes, to celebrate sacred events, and for clothing. The art teacher had students use sticks in the shape of a Y and weave string and yarn around them. This created the loom so the students could interweave the "warp" threads and the "weft" threads together. Students were introduced to weaving vocabulary and then chose from brightly colored yarn to create their woven item. The students collected specimens from nature like long grasses, leaves, and flowers to add into their weave. They sat on the grass and on benches working on the weave. It was a relaxing activity and they were so proud to share their work. A basic loom can also be created with cardboard, and natural items such as grasses and twigs can be woven into it.

Weaving and other fiber arts such as spinning and knitting provide a wonderful opportunity to connect to the community by inviting guest grandparents, and/or artists to share their craft. Students can also interview family members at home about their childhood.

Photo 9.2. Weaving with Natural Materials (Helen Corveleyn)

> ### Children's Book Recommendations
>
> - *Where on Earth? Atlas: The World as You've Never Seen It Before* (DK Where on Earth? Atlases)
> - *Timelines of Everything: From Woolly Mammoths to World Wars* (DK Children's Timelines)
> - *Woven of the World* by Katey Howes and Dinara Mirtalipova
> - *Braiding Sweetgrass for Young Adults: Indigenous Wisdom, Scientific Knowledge, and the Teachings of Plants* by Robin Wall Kimmerer, Monique Gray Smith, and Nicole Neidhardt
> - *Abuela's Weave* by Omar S. Castañeda and Enrique O. Sánchez
> - *The Arabic Quilt: An Immigrant Story* by Aya Khalil and Anait Semirdzhyan
> - *Farmer George Plants a Nation* by Peggy Thomas and Layne Johnson
> - *Remember* by Joy Harjo and Michaela Goade
> - *The Rough-Face Girl* by Rafe Martin and David Shannon
> - *Sofia Valdez, Future Prez: A Picture Book (The Questioneers)* by Andrea Beaty and David Roberts
> - *We Are Water Protectors* by Carole Lindstrom and Michaela Goade
> - *The Water Princess* by Susan Verde, Georgie Badiel, and Peter H. Reynolds
> - *Nya's Long Walk: A Step at a Time* by Linda Sue Park and Brian Pinkney
> - *One Plastic Bag: Isatou Ceesay and the Recycling Women of the Gambia* by Miranda Paul and Elizabeth Zunon
> - *The Boy Who Harnessed the Wind* by William Kamkwamba, Bryan Mealer, and Elizabeth Zunon

Inks and dyes

Creating dyes from nature is an ancient and fascinating process. Our human ancestors used color from minerals, plants, shellfish, lichens, and insects to create paint, pottery, cosmetics, and textiles. Natural dyes can be made with household items such as onion skins, avocado, and berries. For older children, there are chemistry connections possible in creating more advanced dyes. All the activities can be easily connected to research on specific colors, cultures, or the history of dyeing techniques (blue and purple, in particular, have such interesting histories!). For example, dyeing fibers can be connected to the study of Ancient Egypt, Europe, Greece, Japan, and of Native American cultures. The dyes can be created to make paints for paper or for dyeing natural fibers such as wool or cotton. Students can also explore the dyes to make prints or patterns.

Here is a selection of dyes that are classroom and child friendly:

- Green: spinach, parsley, black-eyed Susans, green tea, nettles (be careful when fresh—it stings!)
- Yellow: goldenrod, dandelion flowers, onion skins, marigolds
- Orange: turmeric
- Brown: coffee grounds, tea bags
- Pink: avocado skins and pits, beets
- Purple: red cabbage

A mordant (such as ferrous sulfate, aluminum acetate, soy milk) can be added to brighten or deepen colors, however, it is not necessary for the classroom purpose of exploring the process. There are numerous YouTube videos and websites about dyeing with children. The basic steps for dyeing fibers include:

1. Pre-soak the cloth (add mordant, optional).
2. Boil water and pour over plants, or heat plants in water for thirty minutes to three hours. If using cold water, wait twenty-four hours.
3. Strain-out plant materials.
4. Dip or immerse fabric in the liquid.

In addition to cloth, botanical materials can be used to make simple inks for painting and writing on paper. The paper can be made by students as well!

Using Drones to Teach Geography and Mapping

I (Helen) was lucky enough to have co-taught with a wonderful technology coach for a few years at HES. We did a lot of great lessons together using coding websites, building Piper computers, exploring robots, and creating tech lessons. One of my favorite lessons that we developed in the classroom was using drones as a model of real-world applications of technology.

Even small drones get students all fired up! They can be excellent teaching tools to model scenarios that happen in real life. I like to introduce drones to kids by asking them what they know about drones and how they can help in the world. Then with a quick slide deck, I show them pictures of the following categories:

1. conservation (mapping and tracking large groups of wildlife on land or in the sea, illegal poaching, population counts, etc.);
2. organ transport in large cities;
3. geographical mapping of remote areas;
4. environmental engineering (looking at structurally unsound areas like mines and restoration sites);
5. reforestation techniques.

Students are typically fascinated by the wide range of uses of drones. We then turn to actually learning to fly drones in the classroom: I found small drones to use that have a Blockly coding system on them and their own individual Wi-Fi. The coding program is compatible with iPads, Chromebooks, and iPod touch devices. Once students know how to code them, we set up different scenarios for them to complete. They may have to code the drone to fly to one area of an obstacle course to "pick up goods." The goods are usually written on a small Post-it Note they attach to the drone and then deliver to other areas of the room. This could be aid in a natural disaster caused by climate change; it could be tracking large pods of whales in the ocean; or it could be a medical drop-off during a weather-related catastrophe. The drones make it fun, the coding makes it relevant, and the scenario fits the DCI or the topic you are teaching. Think outside the box for this one, because the kids will be asking for more—it's a guarantee.

Authentic Advocacy Projects

Civic engagement and advocacy are at the heart of social studies. When given the opportunity, students enjoy actively participating in society through advocacy and taking action. Elevating "student voice" is a key component to advocacy at an early age. The most important part of advocacy is that it should happen in the classroom organically. When ideas are teacher driven, student buy-in may not develop into passion for a cause. While they may see the importance in advocating for a cause, if they do not initiate the ideas and planning, the results may not yield a lasting impact on student learning.

There are ways for teachers to authentically guide students to find their passion. For example, if students are studying ocean plastics and their impact on the world's oceans, it helps for them to begin by looking at a variety of pictures, articles, and videos organized in a choice board so that students can curate a self-guided exploration of the topic. When the teacher organizes a large variety of mixed media for students to choose from, it helps students feel they have the choice to frame their background knowledge. From there, each student gets to know what they are interested in for a class discussion. For example, a student may be really into whales, consumes every article and video on wildlife that the teacher has found, and is the wildlife expert for the group discussion. Another may read all about the economic impacts that human pollution has on the oceans. The diversity of interests promotes deeper conversations and leads to greater buy-in to a large topic. This is where action is "born."

The Importance of Audience

As Grant Wiggins and Jay McTighe emphasize in the *Understanding by Design* approach, children benefit from authentic audiences and opportunities to present professional opinions to adults outside of their school communities. One of the ways to do this is by gathering a panel of experts for students to present their ideas to. For example, students may investigate environmental problems such as litter in the waterways and come up with the idea of banning single-use plastics in a town with a beach, lake, or river. Inviting town council representatives to the school is a powerful way for them to learn about local government. Representatives can provide feedback to students about their ideas and what is feasible in the real world. Or take this a step further: older elementary students could present research and solutions (recommendations) at a town council meeting. Local universities or community groups are other possible audiences for their research. Anyone outside the immediate school community elevates the "student voice" as they recognize the importance of student findings.

Teacher in the Limelight: Kate Atkins

In our teacher spotlight, we highlight Kate Atkins, a teacher in Philadelphia who seamlessly integrates teaching literacy skills and environmental advocacy with her second-grade students.

As I walked to school on Monday morning, I wasn't sure how angry my students would be. Over the weekend a plastics factory had spilled thousands of gallons of toxic chemicals into the Delaware River. Philadelphians had been warned the water might not be safe to drink. I thought the event might register to my students because they had chosen the issue of pollution and its impact on animals as their service/learning topic for the year. We had spent several months learning about how pollution gets into water and hurts aquatic animals.

I was right. Orla walked in holding two plastic water bottles aloft and said, with her ironic smile, "This is just making it worse, Ms. Atkins." Students were shaken by having their grown-ups tell them they could not brush their teeth with tap water and upset they couldn't use their reusable bottles.

We did some quick research together on the Smart Board and found the name of the factory and the name and picture of its owner. A contingent of students, in their no-shades-of-gray second-grade way, suggested that we hurt or kill him.

"He's a human being," countered Simone.

"Violence is never the answer," Olivia stated firmly.

We decided instead to write the CEO of the offending corporation a letter. Conveniently, our writing unit was to compose persuasive letters, so we worked as a group to think of the different kinds of reasons—ethos, logos, pathos—we could use to persuade him. We thought collectively about what an appropriate "apology of action" would be. Some students wanted to shut the factory down entirely or move it away from the river, but the group settled on a rigorous schedule of pipe inspections, a shift to cleaner chemicals, and a hefty fine to support environmental protection and other related causes.

Students volunteered to facilitate small groups writing different paragraphs of the letter, and all who wanted participated in the process. After about thirty minutes of cheerful, authentic writing, the groups delivered their paragraphs. I typed them up over lunch and read the finished letter to the class. Cornelius beetled around the room during independent work time and got a signature from everyone, counting and eventually using a checklist to make sure nobody was missed. Ada, who can write neatly and small, addressed the envelope, which led to a quick teachable moment about where to make line breaks in an address. I told the kids I would walk across the street and mail it after I dismissed them, and as they lined up a number of them asked, "Do you have the letter?" Ada told her mom, "My writing is going to be the first thing [the CEO] sees!"

> Frank Bozich
> President and CEO
> Trinseo LLC
> 1000 Chesterbrook Blvd Ste 300
> Berwyn, PA 19312-1084
>
> Dear Mr. Bozich,
>
> We think it is unfair that you get rich and the fish get to die.
>   We are sad that you get 6,000,000 dollars every year but you do not check the pipes enough to make sure they do not burst.
>   We are sad that I can't drink tap water. We are sad because I want fresh water from the Delaware River. We are sad because the fish are dying because of the plastic spills. The river is polluted.
>   The chemicals that spilled in the water are methyl methacrylate, butyl acrylate and ethyl acrylate. They are all hazardous to humans. Chemicals are so bad for the earth. They are killing animals and people.
>   Our class thinks you should check the pipes so they don't spill in the Delaware River. You should have less harmful chemicals to make your products so when it spills it's not as much of a problem.
>   We also think you should pay a fine to help St. Jude's, help with river cleanup, help the hurt animals, help the World Wildlife Fund, and help the Philadelphia Water Works. You get millions of dollars every year. You have more than enough to pay a really big fine.
>   Please contact our teacher at satkins@philasd.org to set up a meeting to talk about how you can fix the problem.
>
> Sincerely,
>
> The Students of Room 202

During the morning discussion, we had also written an email to the advocate who had recently spoken to our class about plastics in water, Faran Savitz from Penn Environment. We had asked him what we could do to help. The next day the first thing Everly said to me was, "Did Faran write back?" Faran had written back, suggesting that we gather letters to the governor. I wrote on the board, "How many letters will it take to persuade the governor?" and we had a nice quick math and social studies lesson on how to write large numbers and how many people live in Pennsylvania. Students settled on 150 letters as a good initial goal. We started with twenty-seven, which is the number of students in our class. Then everyone raised the number of fingers of other people in their households who could sign and we counted around the room chorally by twos and threes and fours, up to about sixty. We then wrote the number of students in the third- and fourth-grade classes, rounded those numbers, and estimated how many letters we would have if most of the students in those classes signed too.

A group of writers volunteered to edit our letter to the CEO to make it into a letter to the governor. During lunch I ran 150 copies and talked to the third- and fourth-grade teachers I could find. One of them invited us in right after lunch, so five fearless students peeled off from our line to make their case and came to class fifteen minutes later smiling triumphantly and waving handfuls of letters.

"The whole class signed!" said Elliah with wide eyes.

"It wasn't so scary since I went to that class from the day you were absent," said Julian proudly.

I brought in a box, cut a long slit in the side, and taped a 200 chart to the top of it. Students who did not want to talk to any of the other classes happily volunteered to "count in" the letters. At first they did it one by one, and then with my gentle encouragement were soon counting out stacks of ten, crossing off all the numbers from, say, sixty-three to seventy-two as they did so. At the end of the day they reported out how many letters we had so far: eighty-nine!

At report card conferences that afternoon, it was obvious the kids had been talking about the project. Elora had already smuggled a letter home and given it to her parents to sign. Another parent, who works in advocacy, agreed to come in the next week to advise the class on how to get the governor to pay attention.

The student I was wondering most about is a kid who has some severe anxieties about pollution getting into her body. I knew this issue and was worried about how this project might affect her. I had tried to keep the conversation focused on how the pollution hurt animals not humans. Her dad said, "She has never had this before." He said the class campaign had been very helpful to her anxiety, giving her some control and agency.

HES Green Team Advocacy Project

In 2022 New Jersey banned single-use plastic bags throughout the state. Just prior to that, individual municipalities started to look into banning the use of single-use plastic bags within the town. Paired with our local watershed organization (The Watershed Institute in Pennington, New Jersey), students from the HES Green Team collaborated on a "learn about trashy plastics" booth at a community event where students educated the school community about single-use plastics. Because of that partnership, The Watershed Institute asked a group of students to "testify" and present at the borough council meeting in support of the bag ban in our community. Since the request to the students for support came from an outside organization, they were far more serious about the message they delivered. The students supported their presentation with facts, dressing appropriately and effectively communicating with adults. They drove the project. They were self-motivated because they felt (and were) valued.

Sometimes past students come back and interview me (Helen) for their high school class graduation project. They ask, "How did you know that setting up a presentation to town council would stick with us as an experience that we will always remember into our adulthood?" My answer is always, "You did the work, which is why it was meaningful. It wasn't a teacher that assigned something to you. You were a true advocate for the voiceless ecosystems you were protecting." They experienced authentic community action and learned that others depended on them to communicate a message of hope for the future.

150   CHAPTER 9

Integrating Conservation in the Curriculum

Conservation is an important global issue, and integrating nature-based learning with social studies provides a platform for students to explore and address conservation challenges. Examining case studies of successful conservation efforts, analyzing the role of government policies and international agreements, and proposing solutions to protect and restore ecosystems are all powerful and meaningful assignments. Students develop a sense of agency and understand the collective responsibility to safeguard the natural world for future generations.

TEACHING WITH HOPE

One of the most important tools when planning conservation lessons is a self-check, which includes questions such as: "Did I add an aspect of hope to my conservation messaging?" If teaching about alternative energy sources and all of the social, economic, and governmental aspects of a topic like offshore wind energy, ask, "Did I integrate success stories?" If not, it is essential to find a short article or video that is a success story, or an innovation to highlight. An excellent example is by Dr. Ayana Elizabeth Johnson, the cofounder of the Urban Ocean Lab and the All We Can Save Project. She is a personal hero of Helen's, and it's worth following, watching, and reading anything she produces. Her work is not only inspirational but also accessible for students. An excellent resource to check out for infusing hope into action is her Climate Action Venn Diagrams (Figure 9.1). This is an amazing tool to connect kids

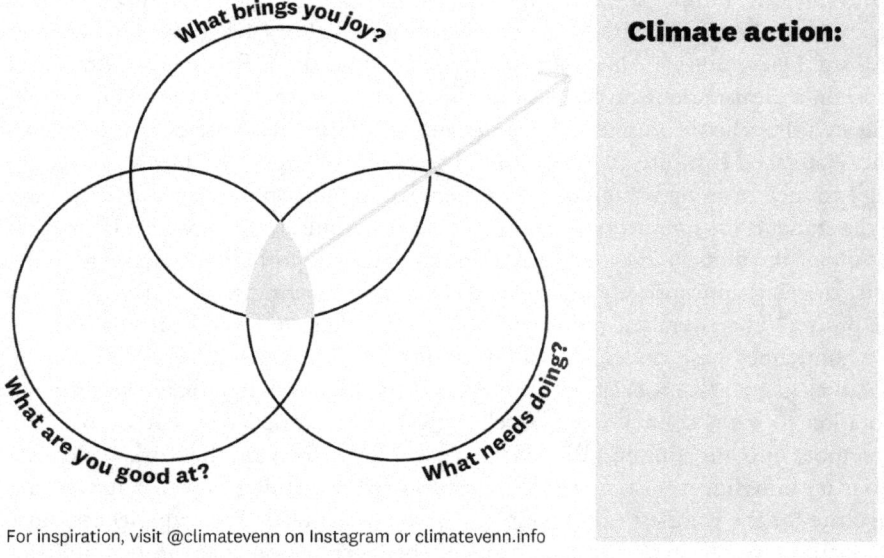

For inspiration, visit @climatevenn on Instagram or climatevenn.info

Figure 9.1. Climate Action Venn Diagrams (Copyright © Dr. Ayana Elizabeth Johnson)

to their own passions and joys and use those positive aspects of their personalities to apply to positive action items for our climate. It is a great integration with SEL. Pairing students on this task results in them complimenting each other on their strengths and reaps some truly insightful and innovative ideas for climate action from each student.

Environmental degradation is a powerful and overwhelming concept. Students of any age are going to be deeply influenced by the presentation of the material and the gravity of the problems we face as a society. As educators we have a responsibility to provide the framework for action. Even if it's a small project, action reduces anxiety! This is an important message for students. The strategy of teaching a lesson and then having students brainstorm a positive action for the environment—which they can either put into action immediately or create long-term goals for—provides a sense of agency for the students.

GETTING INVOLVED IN COMMUNITY ACTION PROJECTS

When considering how to get involved in local environmental projects, the access points might vary. Sometimes connecting with parents is a good starting point—after all, they have a vested interest in the class if their child is in it! There could be parents in the class who work for the state Environmental Protection Agency or are a board member of a local community conservation group. Other institutions and groups to consider include: local universities, nonprofits, outdoor education programs, camps, farms, or conservation associations. They may have small projects that a class can participate in, such as helping to pack seeds for a community garden, or planting native plants in a nearby neighborhood. They may need supplies or assistance with a drive to collect needed goods. All of these service opportunities help to create a relationship with a group that promotes action and advocacy. After students participate in these activities, it is important they reflect through a "do now" or an "exit ticket" and/or post about it on social media. Reporting out how kids are using their voices for good puts positive energy into the world, and the students love to see the responses they get from people. This could be through social media, a local newspaper, or even your district newsletter. This can lead to further community connections and activism opportunities.

THINK GLOBALLY: UN SUSTAINABLE DEVELOPMENT GOALS

A wonderful way to integrate global citizenship into the classroom is to look at the UN Sustainable Development Goals. Based on seventeen areas of focus, the United Nations created this framework to guide both formal and informal educators and NGOs. The UN/UNESCO sustainability goals provide a pathway for addressing global challenges and creating a more sustainable future. Integrating nature-based learning with social studies allows students to explore these goals and understand their significance. Students can investigate how local actions align with the sustainability goals, analyze the impact of human activities on the environment, and propose strategies for action.

> **Teacher Resources**
>
> - Video: Aztec Chinampas, https://www.history.com/videos/aztec-ingenuity
> - Ayana Elizabeth Johnson Projects:
>
>   Website: https://www.ayanaelizabeth.com/
>   The All We Can Save Project: https://www.allwecansave.earth/
>   Urban Ocean Lab: https://www.urbanoceanlab.org/
>   Book: *All We Can Save: Truth, Courage, and Solutions for the Climate Crisis* by Ayana Elizabeth Johnson and Katharine K. Wilkinson.
>
> **Recommended Books**
>
> - *The Brilliant History of Color in Art* by Victoria Finlay
> - *The Art of the Loom: Weaving, Spinning, and Dyeing across the World* by Ann Hecht
> - *The Secret Lives of Color* by Kassia St. Clair
> - *Not Too Late: Changing the Climate Story from Despair to Possibility* by Rebecca Solnit
> - *Saving Us: A Climate Scientist's Case for Hope and Healing in a Divided World* by Katharine Hayhoe
> - *Protest! How People Have Come Together to Change the World* by Alice Haworth-Booth and Emily Haworth-Booth
> - *The Young Activist's Guide to Building a Green Movement and Changing the World: Plan a Campaign, Recruit Supporters, Lobby Politicians, Pass Legislation, Raise Money, Attract Media Attention* by Sharon K. Smith
> - *Advocating for the Environment: How to Gather Your Power and Take Action* by Susan Inches
> - *Braiding Sweetgrass: Indigenous Wisdom, Scientific Knowledge, and the Teachings of Plants* by Robin Wall Kimmerer
> - *Gathering Moss: A Natural and Cultural History of Mosses* by Robin Wall Kimmerer

These goals can be used in classrooms in many different ways. In the grade three through grade five band, one approach could be printing out each of the icons and scattering them around the room, having your students rank their importance. This is not to say that some of the goals are more important. This activity generates incredible discussions within the classroom. Begin by acknowledging that all of the topics are important or the UN would not have chosen them. Then students can respond to some prompts: What goals are the top three that impact our community? Are they different from the top three that affect our state? Our nation? Our world? This gives a concrete way of beginning the conversation about global citizenship in the classroom. Having a "question parking lot" on hand is helpful—a spot for questions as they come up, because they will! After discussing with the class, maybe pick one to focus on as an "actionable" goal for the classroom. This builds an authentic classroom community and gives students a sense of agency. Be prepared to be shocked by the awesomeness that follows! This is the best part of working with children. This type of activity typically fuels a sense of hope in humanity for a long while. Their pure intentions remind us that there is good in our world.

References

INTRODUCTION

American Academy of Child and Adolescent Psychiatry. (2020). *Screen time and children*. https://www.aacap.org/AACAP/Families_and_Youth/Facts_for_Families/FFF-Guide/Children And-Watching-TV-054.aspx

American Academy of Pediatrics Council on Communications and the Media. (2013). Children, adolescents and the media. *Pediatrics, 132*(5), 958–961. doi:10.1542/peds.2013-2656

Charles, C., & Louv, R. (2009). *Children's nature deficit: What we know and don't know*. Children and Nature Network (pp. 1-32). Retrieved from https://drboulet.com/wp-content/uploads/2012/08/CNNEvidence-of-Nature-Deficit.pdf

Dillon, J., Rickinson, M., Teamey, K., Morris, M., Choi, M. Y., Sanders, D., & Benefield, P. (2016). The value of outdoor learning: Evidence from research in the UK and elsewhere. In J. Dillon (Ed.), *Towards a convergence between science and environmental education: The selected works of Justin Dillon* (1st ed., pp. 179–185). Taylor & Francis.

Jordan, C., & Chawla, L. (2019). A coordinated research agenda for nature-based learning. *Frontiers in Psychology, 10*, 766. doi:10.3389/fpsyg.2019.00766. PMID: 31065246; PMCID: PMC6487259.

Kuo, M., Barnes, M., & Jordan, C. (2019). Do experiences with nature promote learning? Converging evidence of a cause-and-effect relationship. *Frontiers in Psychology, 10*, 305. doi:10.3389/fpsyg.2019.00305. PMID: 30873068; PMCID: PMC6401598.

Louv, R. (2008). *Last child in the woods: Saving our children from nature-deficit disorder*. Algonquin Books.

National Recreation and Park Association. (2017). *Children in nature: Improving health by reconnecting youth with the outdoors*. https://www.nrpa.org/uploadedFiles/nrpa.org/Advocacy/Children in-Nature.pdf

Oberle, E., Zeni, M., Munday, F., & Brussoni, M. (2021). Support factors and barriers for outdoor learning in elementary schools: A systemic perspective. *American Journal of Health Education, 52*(5), 251–265.

Spielmaker, D. M., & Leising, J. G. (2013). *National agricultural literacy outcomes*. Logan, UT: Utah State University, School of Applied Sciences & Technology. https://www.agliteracy.org/resources/outcomes/

CHAPTER 1

Abramson, A. (2022, January 1). Children's mental health is in crisis. *Monitor on Psychology*, *53*(1). https://www.apa.org/monitor/2022/01/special-childrens-mental-health

Ballew, M. T., & Omoto, A. M. (2018). Absorption: How nature experiences promote awe and other positive emotions. *Ecopsychology*, *10*(1), 26–35.

Durlak, J. A., Weissberg, R. P., Dymnicki, A. B., Taylor, R. D., & Schellinger, K. B. (2011). The impact of enhancing students' social and emotional learning: A meta-analysis of school-based universal interventions. *Child Development*, *82*, 405–432.

Jo, H., Song, C., & Miyazaki, Y. (2019). Physiological benefits of viewing nature: A systematic review of indoor experiments. *International Journal of Environmental Research and Public Health*, *16*(23), 4739. https://doi.org/10.3390/ijerph16234739

Leeb, R. T., Bitsko, R. H., Radhakrishnan, L., Martinez, P., Njai, R., & Holland, K. M. (2020). Mental health-related emergency department visits among children aged <18 years during the COVID-19 pandemic—United States, January 1–October 17, 2020. *Morbidity and Mortality Weekly Report*, *69*(45), 1675–1680.

Pfeifer, E., Fiedler, H., & Wittmann, M. (2020). Increased relaxation and present orientation after a period of silence in a natural surrounding. *Nordic Journal of Music Therapy*, *29*, 75–92. doi:10.1080/08098131.2019.1642374.

Taylor, R., Oberle, E., Durlak, J. A., & Weissberg, R. P. (2017). Promoting positive youth development through school-based social and emotional learning interventions: A meta-analysis of follow-up effects. *Child Development*, *88*, 1156–1181. doi:10.1111/ cdev.12864

Tillmann, S., Tobin, D., Avison, W., & Gilliland, J. (2018). Mental health benefits of interactions with nature in children and teenagers: A systematic review. *Journal of Epidemiology and Community Health*, *72*(10), 958–966. https://jech.bmj.com/content/jech/72/10/958.full.pdf

CHAPTER 4

Ozer, E. J. (2007). The effects of school gardens on students and schools: Conceptualization and considerations for maximizing healthy development. *Health Education & Behavior*, *34*(6), 846–863.

Spielmaker, D. M., & Leising, J. G. (2013). *National agricultural literacy outcomes*. Utah State University, School of Applied Sciences & Technology. https://www.agliteracy.org/resources/outcomes/

CHAPTER 5

Ozer, E. J. (2007). The effects of school gardens on students and schools: Conceptualization and considerations for maximizing healthy development. *Health Education & Behavior*, *34*(6), 846–863.

CHAPTER 6

Brouwer, R. J. N., & Neelon, S. E. B. (2013). Watch me grow: A garden-based pilot intervention to increase vegetable and fruit intake in preschoolers. *BMC Public Health*, *13*(1), 1–6.

Caraher, M., Seeley, A., Wu, M., & Lloyd, S. (2013). When chefs adopt a school: An evaluation of a cooking intervention in English primary schools. *Appetite, 62*, 50–59.

Corveleyn, H., Ammentorp, L., & Howley, L. (2020). The promise of pickles: Hydroponics in the classroom. *Childhood Education, 96*(6), 14–21. doi: 10.1080/00094056.2020.1846374

Cunningham-Sabo, L., & Lohse, B. (2013). Cooking with kids positively affects fourth graders' vegetable preferences and attitudes and self-efficacy for food and cooking. *Childhood Obesity, 9*(6), 549–556.

Morgan, A. F. (1927). The value of certain home-economics courses as a means of general education. *School Review, 35*(7), 518–529.

Ozer, E. J. (2007). The effects of school gardens on students and schools: Conceptualization and considerations for maximizing healthy development. *Health Education & Behavior, 34*(6), 846–863.

Russell, J. L. (2011). From child's garden to academic press: The role of shifting institutional logics in redefining kindergarten education. *American Educational Research Journal, 48*(2), 236–267.

CHAPTER 8

Fenichel, M., & Schweingruber, H. A. (2010). *Surrounded by science: Learning science in informal environments*. Board on Science Education, Center for Education, Division of Behavioral and Social Sciences and Education. National Academies Press.

NGSS Lead States. (2013). Next Generation Science Standards: For states, by states. National Academies Press. www.nextgenscience.org/next-generation-science-standards.

Rios, J. M., & Brewer, J. (2014). Outdoor education and science achievement. *Applied Environmental Education & Communication, 13*(4), 234–240. doi: 10.1080/1533015X.2015.975084

Index

access, to nature, 1, 17, 59, 134
activities, 101, *121–23*; fibers and, 65–66; in gardens, 54, 56; mindfulness and, 13–15; multisensory, 24–25; supplies for, 38; ten-frame, 111; for vegetables, 89–90, *91*, 92–95
Actual Size (Jenkins), 119
advocacy, 139–45, 150–52; health and, 90; HES Green Team and, 149; language arts and, 25–26; literacy and, 147–49; students and, 146
aeroponic gardens, 71
afterschool programs, 82, 96
Agrawal, Natasha, 78–79, 109–10, *109–10*
agriculture: literacy on, 5, 49, 65; National Center for Agricultural Literacy, 5, 49, *50–52*, 63; STEM and, 139–41
algebra, 112
alphabet, 25
animals, 25, 37, 44–47, 119; ecosystems and, 104; engineering and, 100–101, 104; observation of, *129*; poetry and, 32; resources on, 105; STEM teaching and, 43; therapy dogs, 18–21, *19*; tracks of, 42
apps, 45, 118–19, 126
aquaponics, 69, 81, *81*, 85; collaboration and, 82–83; fish and, 71–72, 82, *82*; preparation for, 70, 80, 82–83; resources for, 86; students and, 83–84, *84*
art: science and, 117; social studies and, 142
art projects, 144–45; collage, 15, 66; mandalas, *14*, 15; observation and, *44*, 44–45; physical science and, 106

Atkins, Kate, 147–49
Aztec floating islands STEM challenge, 139–41, *140*

bees, 31–32, 40, 76
behavior: field trips and, 115, 136; pro-environmental, 3–4, 15–16, 23
biomimicry, NGSS and, 99–100
birds, 119; books on, 45; ecosystems and, 43–44; nests of, 45–46; resources on, 47
Birds STEAM project, HES, 43–46
boat construction, 101, *102–3*
books, 118–19, 144, 152; on birds, 45; on fibers, 64; on food, 64, 89, 95; on gardens, 67; on mindfulness, 22; nature in, 23; for outdoor activities, 39; on plants, 77; poetry and, 36; STEM, 111–12; for teachers, 39, 112; on trees, 42
botany, 126, 134
breath, mindfulness and, 11–12, 18, 29
bridge construction, 101

calm, 11, 16, 22, 66, 97
careers, in STEM, 85, 131
CASEL. *See* Collaborative for Academic, Social, and Emotional Learning Framework
categorization, 38, 56, 65–66, 113, 124
Chawla, Louise, 3
chemistry, 93, 144; hydroponics and, 71–72, 74; soil and, 126, 134
children, 1, 7, 10–11; books for, 23, 42, 45, 64, 77, 89, 112, 144. *See also* students

choice, for students, 18, 24, 90, 131, 146
climate, ecosystems and, 108
climate action, SEL and, 150–51
Climate Action Venn Diagrams, *150*
climate change education, 45, 79–80, 83, 85, 104, 139, 145; hydroponics and, 74
clubs, 61, 69, 82–83
collaboration, 80, 82–83, 131, 134; community and, 63–64, 89, 96, 142; poetry and, 28–29, 31; research and, 44–45; with staff, 85
Collaborative for Academic, Social, and Emotional Learning (CASEL) Framework, 7–9
collage, 15, 66
collection, in nature, 38, 46; creativity and, 106
collection, of data, 56, 83, 85, 110, 113, 134
Common Core ELA Standards, 26–27, 136
Common Core Standards, 111, 121
communication, 147–49; of data, 113, *113*; research and, 136; science and, 121, 134
community, 59, 61, 92; collaboration and, 63–64, 89, 96, 142; projects in, 151; students and, 146, 149
conservation, curriculum and, 150–52
construction, 42; eco-architecture and, 104; found materials and, 14–15, 101, *102–3*, 104; math and, 112; of mobiles, 106; of models, 40, 45, 62, 99–100, 104, 141; observation and, 119; play and, 101, 104
contemplation, SEL and, 7
cooking, 96, *97*, 98; interdisciplinarity and, 87, *88*; social-emotional development and, 87, 97
counting, 111, 143, 149
COVID-19 pandemic, 1, 3, 59, 82–83, 96; mental health and, 10
creativity, 23, 28, 66; collection in nature and, 106; construction and, 104
culinary arts, 6, 87–98
culture: fibers and, 142; food and, 88–89, 92
curiosity, 2, 17, 99; science and, 37
curriculum integration, 4–5, 23, 60–61, 65, 73–77, 115, 117; conservation and, 150–52

data: collection of, 56, 83, 85, 110, 113, 134; communication of, 113, *113*; science and, 56–57
deep-water culture hydroponics (DWC), 71
dehydration, of food, 94–95
demo kitchen, at HES, 87, 96
descriptive writing, 77; prompts for, 24, *24*
design, 42, 46, 74, 99–101, 104, 119, 140–42
Dewey, John, 2
dietary restrictions, students and, 90
different perspectives, 22, 31–32
Dopirak, Kate, *Hurry Up!*, 22
drones, geography and, 145
DWC. *See* deep-water culture
dyes, history and, 144–45

earth science, 107–8
ebb and flow system, of hydroponics, 71–72
eco-architecture, construction and, 104
ecosystems, 40, 42, 62–64, 72; animals and, 104; birds and, 43–44; climate and, 108; fifth-grade unit on, *120–23*, 120–36, *125*, *127–30*, *132–33*, *135*; health of, 124, 131, *132*, 134
education: climate change, 45, 74, 79–80, 83, 85, 104, 139, 145; inquiry-based, 2, 26, 95, 105, 115; observation and, 83; place-based, 126; for teachers, 116; technology and, 104, 119, 142, 145; vocational, 87
Ehlert, Lois, *Growing Vegetable Soup*, 95
Einstein, Albert, 117
emotions, 8–9; poetry and, 28; STEM and, 34; therapy dogs and, 21
empathy, 17, 21, 33, 83
engineering: NGSS and, 101, 104; plants and, 99–100; problem solving and, 140–41
entomology, 126, 134
environmental stewardship, 23, 40, 64; experiential learning and, 139; gardens and, 57–59; pollution and, 26, 139, 146–49; problem solving and, 150–51; water and, 33–34
experiential learning, 2, 66, 80, 82–83, 99; environmental stewardship and, 139; food and, 95; hydroponics and, 71–72; physical science and, 105

INDEX

experimentation, 66, 74, 93, 104–6, 131
expertise, students and, 100, 131, 146
exploration, 42, 74, 146; science and, 134
explorer mindset, observation and, 17
expository writing, research and, 25–26

families of students: NBL and, 34; participation by, 61, 82–83, 89, 149, 151
farms, 49, 65, 69, 74, 92, 141; resources on, 53, 67
farm stands, 94
fibers: activities and, 65–66; books on, 64; history and, 142–45, *143*
field experiences, 6, 115, 117, 120–21, 134, *135*; collaboration and, 131; Common Core ELA Standards and, 136; NGSS and, *121–22*, 124, 126, 131; observation and, 116, 119, 124; resources for, 118; safety and, 136–37; tools and, 126; worksheets for, *125*, *127–30*, *132–33*
field trips, 53, 92; behavior and, 115, 136; preparation for, 131, 134
fieldwork, teachers and, 116, *116*
fifth-grade standards, 26, 93, *121–23*
fifth-grade students, 33, 55, 64, 74, 104, 115, 139–41; ecosystems unit and, *120–23*, 120–36, *125*, *127–30*, *132–33*, *135*; hydroponics and, 80–85
first-grade students, 54, 56, 99, 111
fish, aquaponics and, 71–72, 82, *82*
food, 49, 53, 61, 78–79; books on, 64, 89, 95; culture and, 88–89, 92; experiential learning and, 95; from gardens, 87–88; hydroponics and, 69, 72–73, 80; language arts and, 88–89; life cycles and, 80; math and, 93–94; resources for, 92; safety and, 94–95; smoothies, 90, 96; STEM and, 96; vegetables, 89–90, *91*, 92–95
food preservation, 107; dehydration and, 94–95; NGSS and, 92–93; pickling, *93*, 93–94; social studies and, 92, 94
forces, in physical science, 105–6
found materials, 13, 38, 46, 106; construction and, 14–15, 101, *102–3*, 104
fourth-grade students, 27–34, 40–41, 104, 152
fruit, 94

Gaiman, Neil, "Instructions," 33
games, 32, 56, 126, 131
gardens, *41*, *57–58*, 93; activities in, 54, 56; aeroponic, 71; environmental stewardship and, 57–59; food from, 87–88; of HES, *9*, 49, *55*; history and, 139–45; physical activity and, 15, 54–55; play and, 60; pollinator, 57–58, 62–64, 74, 76–77; preparation for, 57; research and, 141; resources for, 53, 62, 67; science and, 40–41; sensory, 59–60; service and, 62, 72–73, 83; students and, 53–55; summer and, 61; three sisters, 69, 141; weather and, 69, 83. *See also* aquaponics; hydroponics
geography, drones and, 145
geology, 126, 134
grants, 18–19, 27, 34, 45, 82–83, 85, 96
gratitude, 54, 62; observation and, 13–14
Green Team advocacy project, HES, 149
Growing Vegetable Soup (Ehlert), 95
growth mindset, students and, 72, 80
guided meditation, 10, 13; poetry and, 32

habitats, 40, 45, 104
habits, health and, 90
health, 4–5, 7, 16–17, 74; of ecosystems, 124, 131, *132*, 134; food and, 87; habits and, 90; mental, 8, 10, 34–35, 59; mindfulness and, 10–15, 22; SEL and, 8–9; therapy dogs and, 18–21
HES. *See* Hopewell Elementary School
history, 88–89; dyes and, 144–45; fibers and, 142–45, *143*; gardens and, 139–45; STEM and, 139
hope, 150–52
Hopewell Elementary School (HES), 15–16, 26, 37, 42; Birds STEAM project, 43–46; demo kitchen at, 87, 96; garden of, 49, *55*; Green Team advocacy project, 149; hydroponics and, 69, 80–85, *81–82*, *84*; Infinity Garden of, *9*; Salad-brations at, 90, *91*, 92
Hurry Up! (Dopirak and Neal), 22
hydroponics, 5, 93; chemistry and, 71–72, 74; DWC, 71; ebb and flow system of, 71–72; food and, 69, 72–73, 80; HES and, 69, 80–85, *81–82*, *84*; Kratky Method of, 70–71; NFT in, 71; NGSS

and, 71–72, 74, *75–76*; plants and, 71, 72–73, 76–77; preparation for, 70; resources for, 71, 86; SEL and, 73–74, 77–78, 80; students and, 72–73, *73*, 79, *79*, 83; tabletop, 71, 78, *78*
Hydroponics in the Classroom project, 77–79

immersion, 3, 115, 119–20
Infinity Garden, of HES, *9*
innovation, nature and, 99
inquiry-based education, 2, 26, 95, 105, 115
insects, 25, 41; bees, 31–32, 40, 76; observation and, 39–40, *130*
"Instructions" (Gaiman), 33
interdisciplinarity, 2–3, 65; cooking and, 87, *88*; of social studies, 139
investigations, 42, 121; plants and, 74; water and, 105–6; wonder and, 37

Jenkins, Steve, *Actual Size*, 119
Johnson, Ayana Elizabeth, 150–51
Jordan, Cathy, 3
journals, 77, *79*, 115; resources for, 117

kindergarten students, 10, 37, 54, 56, 95, 111; NGSS and, 45, 104
Kratky Method, of hydroponics, 70–71

language arts, 5, 23, 27–36, 79; advocacy and, 25–26; alphabet, 25; descriptive writing and, 24, *24*, 77; expository writing and, 25–26; food and, 88–89; hydroponics and, 77; journals and, 77, *79*, 115, 117
Lanphear, Jeannine, 80–85
Laws, John Muir, 117
life cycles, 41, 73; food and, 80; of plants, 74; seeds and, 54, 56–57
light and shadow, 108, *108*
listening, 25, 28–29, 33, 56, 100, 107
literacy: advocacy and, 147–49; agricultural, 5, 49, 65; therapy dogs and, 18–19, 21
local environments, 5, 131
Louv, Richard, 2–3

mandalas, *14*, 15
Mardell, Ben, 43
mason jars. *See* Kratky Method

math, 56, 66, 113, 148; algebra, 112; art projects and, 106; Common Core Standards and, 111; food and, 93–94
measurement, 66, 108–10, 113
mental health, 8, 59; COVID-19 pandemic and, 10; poetry and, 34–35
mindfulness: activities and, 13–15; books on, 22; breath and, 11–12, 18, 29; cooking and, 87; health and, 10–15, 22; prompts for, 12–13; tea and, 16; teachers and, 10–13; yoga and, 17–18
mobiles, construction of, 106
model construction, 40, 45, 62, 99–100, 104, 141
Montalvo, Naomi, *57*, 57–60
motor skills, 15, 42, 56, 59, 66, 97
multi-language students, 78–79, 109
multisensory activities, 24–25
music, 29, 107

narratives, 24–25
National Center for Agricultural Literacy, 5, 49, *50–52*, 63
National Council for the Social Studies (NCSS), 139
native plants, 58, 62–64, 67, 141–42
naturalists, students as, 118–19
nature, 37, 39–45, 47, 69; access to, 1, 17, 59, 134; in books, 23; collection in, 38, 46, 106; emotions and, 8–9; mindfulness and, 11–15; problem solving and, 99–100; STEM and, 99–113; stress and, 13
nature-based learning (NBL), 2–5, *4*, 11, 16, 25, 80, 150; science and, 115; social studies and, 152; student families and, 34
Nature Harmony Project, 26–36
nature party, 39
NBL. *See* Nature-based learning
NCSS. *See* National Council for the Social Studies
Neal, Christopher Silas, *Hurry Up!*, 22
nests, of birds, 45–46
Next Generation Science Standards (NGSS), 5, 40, 43, 121; biomimicry and, 99–100; engineering and, 101, 104; field experiences and, *121–22*, 124, 126, 131; food preservation and, 92–93; hydroponics and, 71–72, 74, *75–76*;

physical science and, 105–7; Standard for Kindergarten, 45, 104
NFT. *See* Nutrient Film Technique
NGSS. *See* Next Generation Science Standards
No Child Left Behind, 1–2
numbers, 56, 110–13, 148–49
Nutrient Film Technique (NFT), in hydroponics, 71
nutrition, 74

observation, 1–2, 38, 65, 108, *108*, *127*, *129–30*; art projects and, *44*, 44–45; birds and, 46; construction and, 119; education and, 83; explorer mindset and, 17; field experiences and, 116, 119, 124; food and, 93; gratitude and, 13–14; insects and, 39–40; mindfulness and, 12, 22; play and, 43; seeds and, 74; sensory gardens and, 59–60; sketches and, 117; of soil, *128*, *133*; taste and, 95; tools and, 118–19, 134; writing and, 24, 79
organizations: for NBL, 3; for therapy dogs, 19–20
outdoor activities: books for, 39; inquiry and, 105; preparation for, 11–12, 38

pandemic. *See* COVID-19 pandemic
participation, by student families, 61, 82–83, 89, 149, 151
patterns, 13, 39, 66, 111–12
de la Pena, Keeko, 40–41, *41*
performance, 31, 33
permission slips, 20, 38
perspectives, 22, 31–32
phonetics, 24–25
photography, 13, 38, 124; prompts from, 25
physical activity, 17–18, 53, 134; gardens and, 15, 54–55
physical science: art projects and, 106; experiential learning and, 105; forces in, 105–6; NGSS and, 105–7
physics, 106
pickling food, *93*, 93–94
place-based education, 126
plants, 41, 54; books on, 77; engineering and, 99–100; fibers and, 65–66; hydroponics and, 71, 72–73, 76–77; investigations and, 74; native, 58, 62–64, 67, 141–42; observation of, *127*

play, 1, 7; construction and, 101, 104; gardens and, 60; observation and, 43
playgrounds, 37
podcasts, 100–101
poems, by students, 29, *30*, 31–32, 35–36
poetry: books and, 36; collaboration and, 28–29, 31; emotions and, 28; guided meditation and, 32; mental health and, 34–35; Nature Harmony Project and, 26–36; science and, 31, 79; students and, 34–36
policies, school, 57, 72
pollinator gardens, 57–58, 62–64; hydroponics and, 74, 76–77
pollution, 26, 139, 146–49
pre-K students, 56, 80
preparation: for aquaponics, 70, 80, 82–83; field trips and, 131, 134; for gardens, 57; hydroponics and, 70; for outdoor activities, 11–12, 38; pickling and, *93*
preschool students, 10–11, 43, 54, 64, 66, 83
preservation, of food, 92–95, *93*, 107
problem solving, 72, 99–100; engineering and, 140–41; environmental stewardship and, 150–51
pro-environmental behavior, 23; students and, 3–4, 15–16. *See also* environmental stewardship
Progressive Education movement, 2
projects: art, *14*, 15, *44*, 44–45, 66, 106, 144–45; in community, 151; HES Birds STEAM project, 43–46; HES Green Team advocacy project, 149; Hydroponics in the Classroom, 77–79; Nature Harmony, 26–36
project support, 34, 72, 149; for science, 85; from staff, 2, 35
prompts: descriptive writing, 24, *24*; for mindfulness, 12–13; from photography, 25
publicity, 83, 151

quiet, 7, 10–13, 21, 43, 124
quotes, from students, 17, 26, 94, 124, 134

recipes, 88, 94, *97*, 97–98
recycling, reuse and, 40, 41, 62, 94, 147
reflection, 15–16, 77, 151

remote learning, 1, 96
research, 56–57, 99, 131, 134, 144; advocacy and, 147; collaboration and, 44–45; communication and, 136; expository writing and, 25–26; gardens and, 141
resources, 100–101, 151–52; animal, 105; for aquaponics, 86; on birds, 47; farm, 53, 67; for field experiences, 118; food, 92; garden, 53, 62, 67; for hydroponics, 71, 86; for journals, 117; for SEL, 8
respect, 28, 118–19, 136
reuse and recycling, 40, 41, 62, 94, 147
Rousseau, Jean-Jacques, 2

safety, 1, 57, 59, 70, 131; field experiences and, 136–37; food and, 94–95
Salad-brations, at HES, 90, *91*, 92
scavenger hunts, 38
school policies, 57, 72
science, 66, 85; art and, 117; botany, 126, 134; communication and, 121, 134; curiosity and, 37; data and, 56–57; earth science, 107–8; entomology, 126, 134; exploration and, 134; gardens and, 40–41; geology, 126, 134; hydroponics and, 74, 76–77, 83; NBL and, 115; physical, 105–7; poetry and, 31, 79; social studies and, 142; standards for, 139–40
science, technology, engineering, and mathematics (STEM), 1, 6, 99–110, 113; animals and, 43; Aztec floating islands and, 139–41, *140*; books on, 111–12; careers in, 85, 131; emotions and, 34; food and, 96; history and, 139; students and, 85
scientists, students as, 131
second-grade students, 24, 54, 56, 93, 147–49
seeds, 41, 99; food and, 78–79; hydroponics and, 73; life cycles and, 54, 56–57; observation and, 74
SEL. *See* Social-Emotional Learning
self-awareness, 74, 87, 117
self-esteem, 21
self-expression, 30, 34–36
sensory gardens, 59–60
service, 151; gardens and, 62, 72–73, 83; trash collection as, 15–16
sketches and drawings, 38, *44*, 44–45; observations and, 117

Smith, Colby Cedar, *26*, 26–36
smoothies, 90, 96
social-emotional development, 77–78, 117; cooking and, 87, 97
Social-Emotional Learning (SEL), 4–7, 69; climate action and, 150–51; gardens and, 62; health and, 8–9; hydroponics and, 73–74, 77–78, 80; resources for, 8; therapy dogs and, 21
social media, 96
social studies, 6, 66, 148; advocacy and, 139–52; art and, 142; food preservation and, 92, 94; interdisciplinarity of, 139; NBL and, 152; science and, 142
soil, 42, 105, *134*; chemistry and, 126, 134; observation of, *128*, *133*
solar ovens, 107
sound maps, 117
sounds, 24–25, 56, 107
special needs, students with, 15, 100
spelling, 56, 66
squirrels, 43, 119
staff, 62, 83; collaboration with, 85; support from, 2, 35
standards: CASEL Framework, 7–9; Common Core, 111, 121; Common Core ELA, 26–27, 136; fifth-grade, *121–23*; National Center for Agricultural Literacy, 5, 49, *50–52*, 63; science, 139–40. *See also* Next Generation Science Standards; Social-Emotional Learning
STEM. *See* science, technology, engineering, and mathematics
A Stone Sat Still (Wenzel), 22
stress, 21; children and, 10–11; nature and, 13
students: advocacy and, 146; aquaponics and, 83–84, *84*; choice for, 18, 24, 90, 131, 146; community and, 146, 149; dietary restrictions and, 90; environmental stewardship and, 64; expertise and, 100, 131, 146; families of, 34, 61, 82–83, 89, 149, 151; first-grade, 54, 56, 99, 111; fourth-grade, 27–34, 40–41, 104, 152; gardens and, 53–55; growth mindset and, 72, 80; hydroponics and, 72–73, *73*, 79, *79*, 83; kindergarten, 10, 37, 45, 54, 56, 95, 104, 111; mindfulness and, 10–11; multi-language, 78–79, 109; as naturalists, 118–19; poems by, 29, *30*,

31–32, 35–36; poetry and, 34–36; pre-K, 56, 80; preschool, 10–11, 43, 54, 64, 66, 83; pro-environmental behavior and, 3–4, 15–16; quotes from, 17, 26, 94, 124, 134; as scientists, 131; second-grade, 24, 54, 56, 93, 147–49; with special needs, 15, 100; STEM and, 85; third-grade, 54, 94, 109, 142, 148–49. *See also* fifth-grade students
"student voice," 146
summer, gardens and, 61
supplies, for activities, 38
sustainability, 63, 65, 79, 94, 108, 139, 152

tabletop hydroponics, 71, 78, *78*
taste, observation and, 95
taste tests, 89–90, 94
tea, 95; mindfulness and, 16
teachers, 151; books for, 39, 112; education for, 116; fieldwork and, 116, *116*; mindfulness and, 10–13; yoga and, 17–18
technology, 1; education and, 104, 119, 142, 145
ten-frame activities, 111
terrariums, 40
therapy dogs, 18–21, *19*
third-grade students, 54, 94, 109, 142, 148–49
Thompson, Ashley, 18–21
three sisters garden, 69, 141
tools, 38; field experiences and, 126; observation and, 118–19, 134

tracks, of animals, 42
trash collection, as service, 15–16
trees, 29–30, 39, 141–42; books on, 42

United Nations (UN), Sustainable Development Goals of, 152

vegetables, activities for, 89–90, *91*, 92–95
vocabulary, 77, 110, *110*
vocational education, 87

Washington, George, 141–42
water, 107; environmental stewardship and, 33–34; investigations and, 105–6
water cycle, 40, 42
weather, 2, 104, 108, *110*, 113; gardens and, 69, 83
weather station, *109*, 109–10
weaving, 142, *143*
Wenzel, Brendan, *A Stone Sat Still*, 22
wonder, investigations and, 37
word problems, 112
Wordsworth, William, 1
worksheets, for field experiences, *125*, *127–30*, *132–33*
writing: descriptive, 24, *24*, 77; expository, 25–26; observation and, 24, 79

yoga, mindfulness and, 17–18

zoology, 126

www.ingramcontent.com/pod-product-compliance
Lightning Source LLC
Chambersburg PA
CBHW082142230426
43672CB00016B/2934